CONTENTS

PREFACE

Population growth, increasing living standards, and rapidly changing climate have resulted in an increasing demand for freshwater, accelerating the water degradation challenges. There is a compelling need to minimize water consumption and develop approaches to effectively manage existing water resources. On a positive note, water resource management strategies discussed in this book present innovative ways to conserve both quality and quantity. Chapter 1 discusses decentralized water management approaches for intervening the urban water cycle to minimize the environmental and socioeconomic impacts. This chapter concludes with a need to use a suite of tools based on decision support systems for managing urban water resources. Chapter 2 discusses the need for assessing suitability of various types of models for a specific scenario based on the required level of complexity. This chapter discusses in detail the underlying criteria behind model selection, validation, and uncertainty analysis.

Urban watersheds can be more challenging compared to natural watersheds. The urban watersheds include parking lots, roads, and developed structures, all of which contribute to a myriad of anthropogenic pollutants through stormwater runoff. Computer-based models can be used to study water quality issues and to develop a plan to manage watershed level resources. Chapter 3 compares pros and cons of the state-of-the-art watershed models used for managing water resources. Numerical simulations can be performed to compare the current and future water quality scenarios of a given watershed and to estimate the impact of potential water resource management strategies. Chapter 4 presents a case study of an urban region in Hanoi, Vietnam. Water evaluation and planning simulation tool was used to predict the trends and drivers of wastewater generation.

Considering rapidly changing climate and associated weather impacts, it is critical to secure water resources in addition to dealing with the water quality issues. Chapter 5 suggests that climate change models and watershed and precipitation models should be jointly used in order to capture uncertainties in ecological functions, energy and food production and water supply sources. Chapter 6 presents a water use estimation and management tool that examines the effect of climate change and drought conditions on water supplies to ensure adequate buffalo forage. Sustaining both buffalo forage and water supplies during drought conditions requires preparedness and adaptation in response to unfavorable conditions. Finally, water reuse can alleviate the stress on available water resources. For example, effluents from wastewater treatment plants and desalination plants can be treated and reused for managing water crisis. Chapter 7 emphasizes that it is critical to optimize both economical and sustainability parameters during treatment of wastewater effluents and desalination concentrate. In certain cases, valuable metals can be recovered from the concentrate.

Editors
Veera Gnaneswar Gude
Venkataramana Gadhamshetty
Ramanitharan Kandiah

In: Sustainable Water
Editors: V. Gnaneswar Gude et al.
ISBN: 978-1-53617-335-2

Chapter 1

INTEGRATED APPROACHES TOWARD SUSTAINABLE URBAN WATER RESOURCES MANAGEMENT

Seneshaw Tsegaye[1,*], PhD, Jong-Yeop Kim[1], PhD and Kristopher C. Gallagher[2]

[1]Department of Environmental and Civil Engineering,
U. A. Whitaker College of Engineering, Florida Gulf Coast University
Fort Myers, Florida, US
[2]Dewberry Engineers, Inc., Geospatial and Technology Services,
Tampa, Florida, US

ABSTRACT

Global human population growth, a driver for living-standard improvement and rapid economic development, are considered the major triggers that increase challenges for conventional water resource management options. The current urban water management approach does not permit adaptability to address issues such as water scarcity or resilience

---

* Corresponding Author's Email: jkim@fgcu.edu.

to extreme weather events within its design. To ensure a more sustainable future, there is a need for a paradigm shift. Advances in integrated urban water management (IUWM) provide the potential to satisfy the water needs of communities at the lowest cost while minimizing adverse environmental and social impacts. The application of this principle fosters a more modular, decentralized approach to urban water management as a clustered growth approach to urban development, minimizing the distance from source to consumer. Recognizing the synergies and co-benefits through mutually-supportive, cross-sector adaptations is particularly crucial for long-term success. Therefore, sustainable water management approaches that offer a more diverse and versatile set of options for dealing with complex urban water challenges are presented in this chapter. These include the IUWM approach which offers interventions across the entire urban water cycle, a decentralized water management approach that enhances the management of water and wastewater close to their sources, and a water sensitive urban design (WSUD) that integrates the urban water cycle into the overall urban design to minimize environmental degradation while improving aesthetic and recreational appeal. With heavy incorporation into the existing urban environment, the application of geographic information systems (GIS) for water resource management is also discussed which significantly broadens opportunities for resource managers, helps develop decision support systems (DSS), illustrates solutions from predictive models, and helps visualize spatially-correlated data.

INTRODUCTION

It is widely accepted that one of the major global challenges of the 21st century is the lack of access to safe drinking water and basic sanitation. A safe water supply and reliable sanitation are fundamental to a community's health and development. However, providing adequate water supply and sanitation, particularly in urban areas, is a challenge in many countries worldwide (Gleick, 2002). The statistics are overwhelming. Over 750 million people across the globe lack access to improved water sources and more than 2.6 billion people lack access to basic sanitation. Of these numbers, the vast majority live in developing countries.

In order to cope with these challenges, cities must manage their water in a sustainable manner. According to Hill (2016) and Cosgrove & Loucks (2015), sustainable water management includes efficiently providing safe, reliable, and easily-accessible water to residents and businesses while also providing reliable access to sanitation and protecting waterways from pollution. This also requires resilience and adaptability to extreme weather events and climate change that contribute to extremes such as flooding and scarcity. Failure to address these challenges creates a myriad of potential health risks and often prevents these underprivileged residents from contributing to the urban economy.

This chapter begins by considering a wide range of water-related issues that cities are facing. It discusses the challenges of current water management approaches and presents alternative approaches to conventional water management. These include Integrated Urban Water Management (IUWM), decentralized water systems, and Water Sensitive Urban Design (WSUD). The chapter then discusses the use of Geographic Information Systems (GIS) and how GIS can be applied in the planning and design of sustainable water management options. The chapter concludes by highlighting the need for efficient communication by urban planners, environmentalist, urban foresters, landscape planners, engineers, and other professionals to create the necessary paradigm shift.

WATER RESOURCE CHALLENGES IN CITIES

Today, cities across the world are facing a wide range of challenges due to dynamic regional and global pressures that impact the way urban water systems are managed. These global pressures include population growth, urbanization, climate change, and aging infrastructure.

In conjunction with diminishing water resources, these additional pressures shall create future burdens on cities in both developed and developing countries where attempts to efficiently provide consistent resources to their ever-growing populations will struggle, if not collapse. Some of the cities which are facing water related challenges are highlighted in this section.

Lima, Peru is home to more than 8 million people. It is one of the driest cities in the world, receiving only 9 mm of precipitation each year and therefore remains highly dependent on surface water traveling down from the Andes mountains. This water supply varies seasonally, and variations are expected to be intensified by climate change. Added stressors of mistreated wastewater impact the surface water within the city. Only about 9% of the wastewater is properly treated. Given these limitations, Lima has applied a multi-stakeholder engagement to influence national legislation for wastewater reuse.

Water demand in the city of Nairobi, Kenya has been steadily increasing. With the added stressor of periodic and intense droughts, water demand continued increased from 203,000 m^3/d to 579,000 m^3/d from 1985 to 2010 (AWSB, 2012). Furthermore, the city is expected to grow from 3.14 million in 2012 to about 6.4 million by 2035. This will inherently increase pressures on its limited water resources. Nairobi continues to work with local and global entities to develop a municipal master plan that will largely focus on improved water resource management (Eckart et al., 2012).

Singapore, a water-scarce country, depends heavily on imported water from Malaysia. Per agreements signed back in 1962, Singapore may import water from Malaysia until the year 2061. However, extension of this transboundary negotiation beyond 2061 has not been encouraging. The two countries are still far apart in terms of their national requirements. Malaysia is reportedly increasing the price of imported water 20-fold. Due to this, Singapore created alternative options for water resource management. They are working to become more self-sufficient and reduce their dependence on Malaysian imports (Long, 2001).

South Korea is known for its high population density, more than 10 times the global average. The country maintains a population of about 51.5 million and has the 11th highest GDP in the world (IMF, 2019). South Korea's extraordinary economic success story has co-existed with one of the world's most significant water challenges. This includes uncertainties about climate change (flooding and drought) largely from seasonal variation in precipitation, extensive and recursive pollution of fresh water sources (Smith, 2019), marginal renewable freshwater resources per person, and geopolitical tensions with North Korea. Furthermore, fresh water makes up only 100,000 square kilometers (about 3%) of the total territory. In spite of these challenges, access to safe drinking water as well as access to improved sanitation are nearly universal in South Korea (Choi et al., 2017).

It is expected that with increasing global and local change pressures, the business-as-usual approach will strain the availability of water and sanitation services. In developing countries where water is severely scarce, we are beginning to see municipalities broaden their perspectives beyond the well-known conventional water management approaches.

To ensure a more sustainable future, there is a need for more drastic measures—a paradigm shift. This paradigm shift should be based on key concepts including:

1. interventions over the entire urban water cycle, reconsideration of the way water is used (and reused) and a greater application of natural systems for water and wastewater treatment including constructed wetlands and soil aquifer treatment (Tsegaye et al., 2012);
2. a decentralized water management approach that enhances the management of water, wastewater, and stormwater close to their sources; and
3. WSUD that integrates the urban water cycle, into urban design to minimize environmental degradation and improve aesthetic and recreational appeal (Smaniotto et al., 2014).

Through coordinated and flexible planning among water-using sectors, these approaches offer a more diverse and versatile set of options for dealing with larger and more complex urban water challenges.

Conventional Water Resource Management

The current municipal water network servicing most cities today is based on 19th century design principles intended to serve smaller populations from abundant water sources (Zhou et al., 2009). Urban water issues often remain disconnected from broader urban planning processes and basin-level management. Urban master plans have not accounted for the various infrastructural components of urban water management including water supply, wastewater, non-waterborne sanitation, stormwater drainage and solid waste management. Although water supply, sanitation and urban settlement forecasting may be incorporated into basin-wide management plans, these often neglect to acknowledge the cross-scale interdependencies in freshwater, wastewater, flood control, and stormwater (Parkinson et al., 2010).

The current urban water management system in developed countries has managed to provide required services to residents. Institutions are well established and adequate financial and human resources are available. However, the conventional approach does not permit adaptability within its design. When extending past the everyday demand and modeled under extreme conditions, the current design is unable to address issues such as water scarcity or prolonged drought. In the case of developing countries where institutions are weak, such challenges are intensified. If current management solutions and technological interventions are not seriously reformed, the amalgamation of growing populations and diminishing water resources in conjunction with external stressors such as climate change and widespread disease will inflict serious global impacts.

To exacerbate these challenges, the conventional urban water management model in both developed and developing countries has failed to identify uses for the three different urban water streams: potable water, wastewater and stormwater. As a result, high-quality potable water is used in places where non-potable or reclaimed water could suffice (van der Steen & Howe, 2009). As demand increases and resources grow scarce, current models of urban water management and their corresponding infrastructure are on the verge of collapse from the perspective of cost-effectiveness, performance, and sustainability. Moreover, bureaucratic disconnects among discrete governing agencies further mutes water use and reuse opportunities. As urban governments become more complex and specialized, sectoral integration within and between different government entities becomes increasingly important.

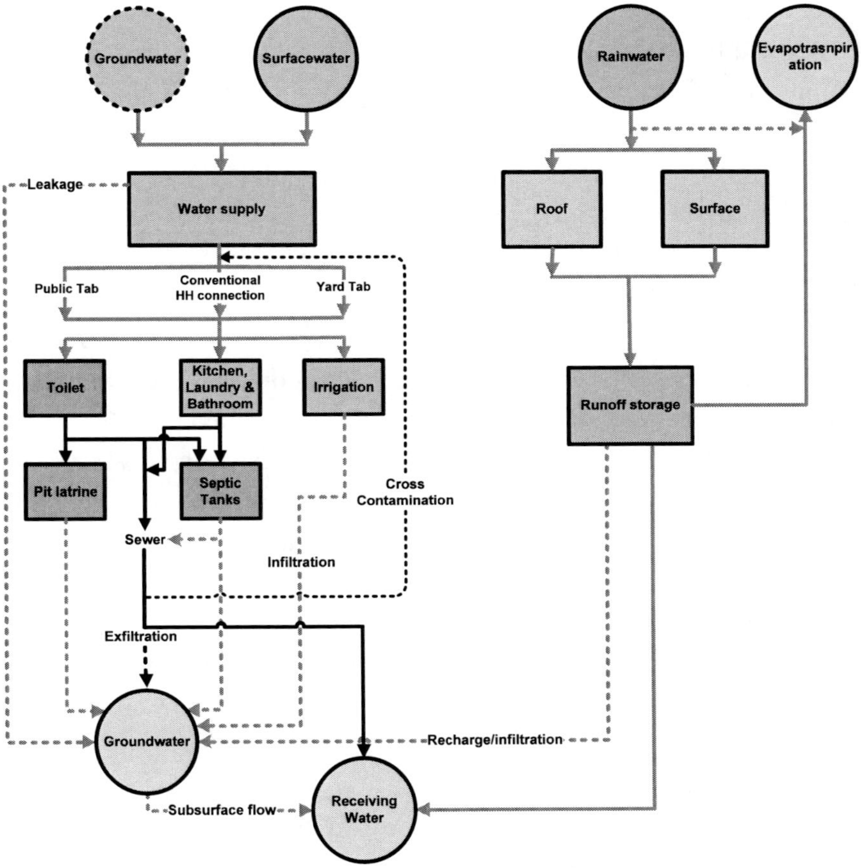

Figure 1. Conventional urban water management.

Building on the bureaucratic gap between policies and agencies, there is a larger gap between urban development objectives and basin-scale management. Urban master plans for individual municipalities have not yet accounted for the larger, basin-scale components of urban water management that impact neighboring cities within the same watershed. Even when well acknowledged by individual municipalities, the transboundary impacts of these components—water supply, wastewater, non-waterborne sanitation, stormwater drainage, and solid waste management—are often neglected (Parkinson et al., 2010). In short, water concerns know no political boundaries. Identifying the disconnect here is simple: the management of water supply, sanitation, and stormwater has not occurred in concert.

Instead, each has been planned and delivered as an isolated service. Figure 1 illustrates this succinctly.

Figure 1 illustrates the interconnections among problems and potential solutions which are often missed. Potable water fulfills all demands, including use for toilets and irrigation where high-quality water is not required. Stormwater systems are designed based on conventional principles that include a collect-redirect-discharge schema. These systems do not enhance the ability to capture stormwater at the source for reuse. As populations grow, urbanization sprawls and industrial activities expand and create dramatic responses in water use, wastewater discharge and flooding.

External stressors not directly related to the water supply also indirectly diminish available water supplies. Climate change is another global pressure that is predicted to create challenges. Significant changes in precipitation patterns will drastically affect the availability and distribution of water. The current 19th-century-based urban water infrastructure is designed as an energy-intensive and rigid system that lacks the flexibility to adapt to the sudden onset of extreme conditions. Given the change pressures just discussed, cities of the future will undoubtedly experience difficulties when attempting to efficiently manage water resources and protect public health in a sustainable manner.

It is not possible to address the current and future challenges of water management using the conventional approach widely implemented today. If cities plan to provide safe and consistent services to their populations, urban water management must consider the intimate integration of the three different and independent components of urban water cycle: potable water, wastewater, and stormwater. The traditional reactive approach when repairing systematic fractures in the current management regime will not be sufficient to cope with these challenges. There is a need for a fundamental change in the way we manage urban water based on a foundation of research, technology and innovation.

In order to address these challenges, the new paradigm of Integrated Urban Water Management (IUWM) needs to be discussed (Mitchel et al., 2004; Jacobsen et al., 2012). The IUWM approach has advanced sufficiently and provides the potential to satisfy the water needs of communities at the lowest cost while minimizing adverse environmental and social impacts.

INTEGRATED URBAN WATER MANAGEMENT (IUWM)

Efficient management of urban water resources as supplies approach scarcity will require integrated solutions across scales (household, neighborhood, city, and catchment), domains (economic, social, and environmental), and institutions (government, private sector, and civil society). The Integrated Urban Water Management (IUWM) approach implements interventions over the entire urban water cycle and reconsiders the way water is used (and reused). IUWM provides an alternative to the conventional approach for an effective and efficient management of scarce water resources by integrating natural systems with industrial processes for a more wholistic approach to water and wastewater treatment.

There is an abundance of positive support for IUWM in the literature. Jacobsen et al., (2012) states that "Integrated urban water management (IUWM) seeks to develop efficient, flexible urban water systems by adopting a holistic view of all components of the urban water cycle (water supply, sanitation, storm water management) in the context of the wider watershed." Additionally, Mitchell (2004) describes IUWM as a new approach for urban water management that takes a comprehensive perspective to urban water services, viewing water supply, wastewater and storm water as components of an integrated physical system recognizing that the physical system sits within an organizational framework and a broader natural landscape.

IUWM seeks to provide sustainable solutions that can respond to the increasing uncertainty about future conditions created by climate change, rapid growth, and other external pressures (Tsegaye et al., 2009; Huang, 2010). Rapidly expanding cities in developing countries are particularly suited to IUWM solutions because new infrastructure and management frameworks can be designed from the scratch using IUWM principles.

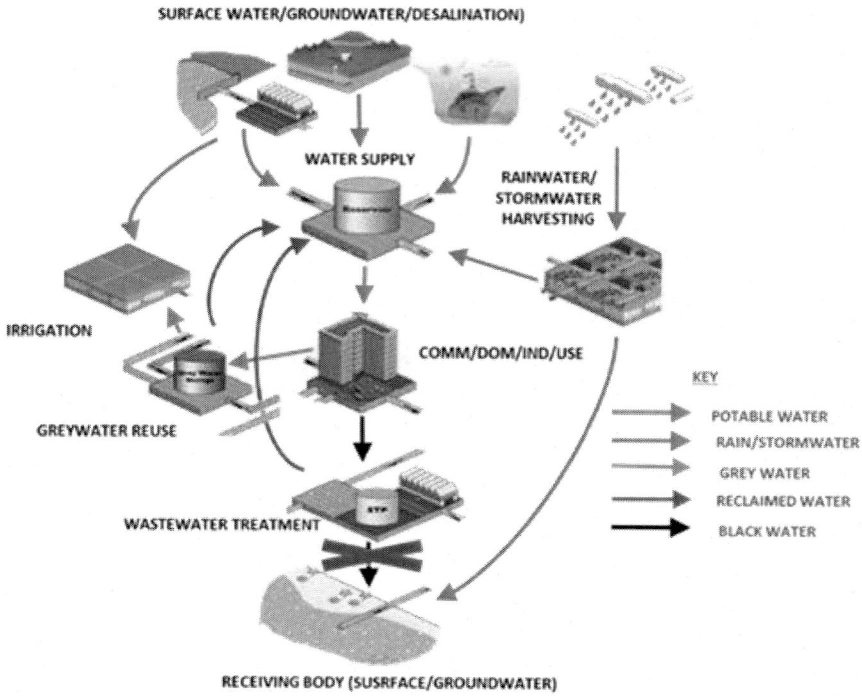

Figure 2. The urban water nexus (Vairavamoorthy et al., 2011).

To implement IUWM, there is an essential need for a new paradigm in urban planning. It is important that water professionals understand and appreciate the significant role of urban planning in potentially supporting the optimization of their water systems. The importance of urban planning to water management requires close coordination early in the development stage between planners and water professionals (Brown et al., 2008). Carefully selected systems can enhance stormwater management and allow reuse of treated wastewater for environmental uses while simultaneously enhancing landscape design.

Critical to the success of IUWM is the early and continuous integration of all stakeholders in the planning and implementation process. The stakeholder participatory process is required to understand the priorities of potential users. It intends that all stakeholders look at the urban water system with common objectives and facilitate the development of innovative solutions to optimize the whole urban water system (van der Steen & Howe,

2009). Both the challenges and solutions stemming from IUWM cross geographic and system boundaries affect mandates between institutions. Therefore, IUWM can only be implemented if institutions agree to work together. In order to execute this holistically, the entire water nexus as a whole must be considered in order to contextualize the local water demand within an integrated framework (Figure 2). This will broaden perspectives for an understanding of the various components of the local system and its impacts on the greater nexus.

When considering water demand, it is important to match water quality for its intended use. The application of this principle offers alternative water sources to be safely used for different purposes. Reclaimed water can be reused for toilet-flushing, gardening or for specific industrial processes. Stormwater harvesting provides a tremendous opportunity to support water conservation among both residential and commercial properties (Jacobsen et al., 2012). The most recent example of South Korea's IUWM strategy can be seen at work on Kido Island, one of the thousands of tiny islands that dot the southwest coast of Korea. The installation of rainwater harvesting tanks will capture scarce rainfall and create options as a potable water source. As of 2011, 59 cities throughout Korea have become “rain cities” including the capital, Seoul, and the major cities of Incheon, Kwangju, Busan, Daejeon and Daegu (Water World, 2012). The state-run Korea Water Resources Corp. (K-Water) has deployed a wide range of IUWM technologies that leverage Korea’s advanced Information and Communication Technologies’ (ICT) infrastructure to minimize loss and optimize performance across the entire water cycle. K-Water is now actively exporting these solutions to other countries (World Water Council, 2015).

The diversity of water profiles creates a portfolio of water sources that expand options beyond surface and groundwater sources and open up opportunities for non-conventional options such as rainwater, greywater and blackwater. For example, an approach to identify sources of cross-contamination from leaking sewers and foul waterbodies can be reflected positively as an opportunity for considering a portfolio of low-quality water sources for reuse in non-potable applications (Mitchell, 2004; Vairavamoorthy et al., 2007; van der Steen & Howe, 2009). Nairobi, Kenya presents a classic case of how the gap between water supply and demand might grow and how IUWM could achieve water security through diversification. An IUWM approach was applied to evaluate the water balance between demand and supply in Nairobi. The analysis shows that non-conventional water sources are viable options to address supply-demand gaps (see Figure 3).

By employing innovative treatment technologies to remedy the aforementioned concerns could open additional opportunities for energy and nutrients harvesting (Bieker et al., 2010; Cornel et al., 2011). The potential of cascading water usage from highest- to lowest-quality needs is the key component behind this approach, allowing for water security through diversity (Alcamo et al., 2008, Gleick, 2009).

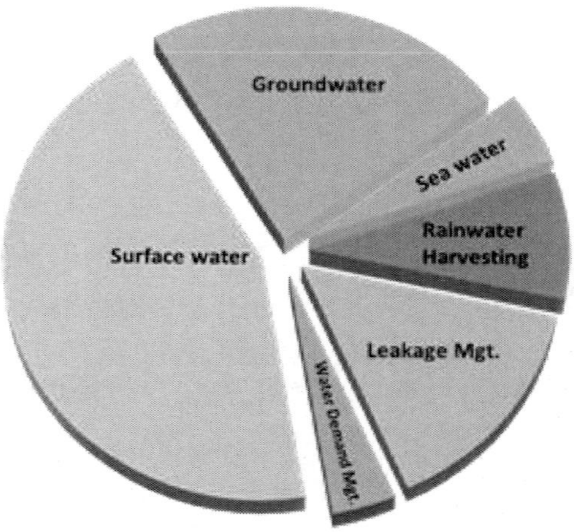

Figure 3. Diversified water sources.

As with all management approaches, it is important to recognize uncertainty that exists within the future change pressures. IUWM calls on the need for flexible systems that have the ability to cope with uncertainties and hence have the capability to adapt to new, different, or changing requirements. The application of this principle fosters a more modular, decentralized approach to urban water management and a clustered growth approach to urban development (Bieker et al., 2010). This decentralized/ cluster approach to urban development optimizes the adaptive capacity of the emerging urban space by allowing infrastructure provision to be staged in a way that traces the urban growth trajectory more carefully.

DECENTRALIZED WATER MANAGEMENT

With increasing global change pressures, it has become obvious that the current practice of urban water management is not sustainable. In order to cope with these challenges, different authors have proposed decentralization of urban water systems. While there is great consensus on the benefits of decentralized water systems; one of the challenges for cities to make this transition is a lack of clear guidance toward transitioning to clustered water systems.

Decentralized water systems are small sub-systems (clusters) that have a large degree of autonomy. According to Bieker et al., (2010), future urban water systems are likely to be more decentralized than conventional systems because water reuse requires reducing the distance between water users and treatment locations. Consequently, closer proximity minimizes energy demand and infrastructure costs and maximizes the recovery of heat energy (Cornel et al., 2011; Newman, 2001; Bieker et al., 2010; Chen and Wang, 2009; and Verstraete et al., 2011).

In addition to efficiently managing intermittent water resources, clustered systems are better adapted to future changes with low effort and without affecting the performance of the entire system (Böhm et al., 2011). Their modular diversity exponentially increases the number of possible configurations that can be achieved for urban water systems from a given set of inputs. Decentralized water systems can be implemented in an incremental fashion, which reduces investment costs and increases facility management efficiency (Tsegaye et al., 2012; Wang et al., 2008; Weber et al., 2007).

In addition, the small-scale benefits of decentralized water systems allow a staged development that traces the urban growth trajectory more closely; implementation can occur more quickly when compared to the conventional approach. According to Wang et al., (2008), the gradual stepwise development of decentralized systems enables the expansion of urban water systems that follows spatial growth, inherently embedding flexibility to municipal water management systems. Clustered systems also provide a better capacity to reduce the risk of contamination, whether through biological or chemical ingression or malicious attacks, due to the small, independent units. Any effect associated with water contamination will be contained within a cluster. Table 1 shows the main features of decentralized and centralized water management approaches.

Table 1. Main difference between centralized and decentralized water management (Domènech et al., 2011)

Factor	Centralized water management	Decentralized water management
Scale	Large-scale systems	Small-scale systems
Types of water sources	Distant and local water sources	Local water sources
Governance	Top-down governance model	Multilevel governance model
Participation	Limited public participation in water management	Active public participation in water management
Awareness	Citizens are alienated from the water cycle	Citizens are more aware of the water cycle
Cost sharing	Highly subsidized	Full cost recovery
Water quality	Very high water quality for all uses	Different water qualities and fit-for-purpose water use
Environmental impacts	Environmental impacts are significant	Environmental impacts are reduced
Resilience capacity	Limited capacity to adapt to extreme situations	Enhanced capacity to adapt to different situations

The implementation of decentralized urban water systems focuses on finding the most effective and efficient balance between rapid urban growth and provision of reliable and sustainable infrastructure systems. Intra-urban recycling, especially water reuse and waste-to-energy recovery, require compact systems which avoid long-distance transports of wastewater. These decentralized hubs foster a close-by treatment approach at the neighborhood or community scale (Daigger, 2009; Bieker et al., 2010).

For example, waste water recycling is one of the most reliable options to reduce urban water demand and provide environmental benefits through waste reduction and energy harvesting. Although the use of both centralized and decentralized recycling systems are assessed based on available site criteria, economic conditions, and facility treatment options, the decentralized systems should be given priority. These smaller hubs diffuse the burden of grey water recycling on a greater centralized system and save on the transportation costs by requiring water and wastewater to travel shorter distances (Ahmed and Arora, 2012; Leas et al., 2014). Figure 4 shows a schematic example of the layout of a decentralized verses a centralized approach for recycling wastewater.

As populations grow, urban areas densify, and demand for clean water rises, we inherently see an evolution in conventional water management approaches. Among the evolution, we see records in the literature of higher levels of energy recovery, new options for nutrient harvesting, and more efficient wastewater reuse. A variety of factors affect the selection of an optimum approach for a particular urban area including local hydrology, available water supplies, water demands, local energy and nutrient-management situations, existing infrastructure, and utility governance structure (Daigger, 2009). A proper approach to economic analysis is critical to determine the most sustainable solutions. The main factor here being an economic one.

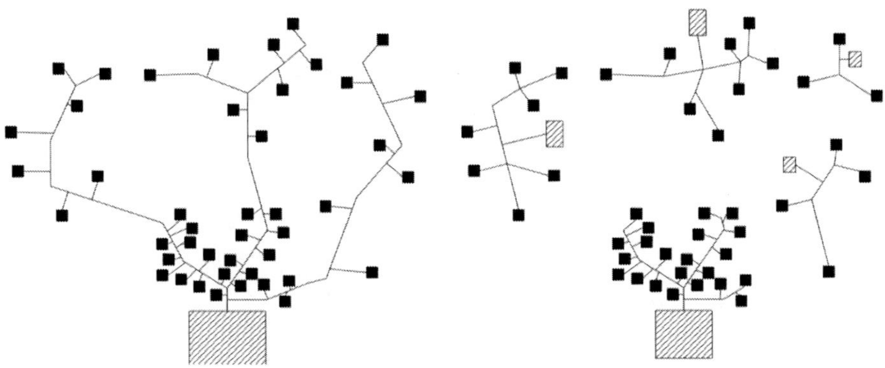

Figure 4. Centralized (left) and decentralized (right) approaches (Nam, 2006).

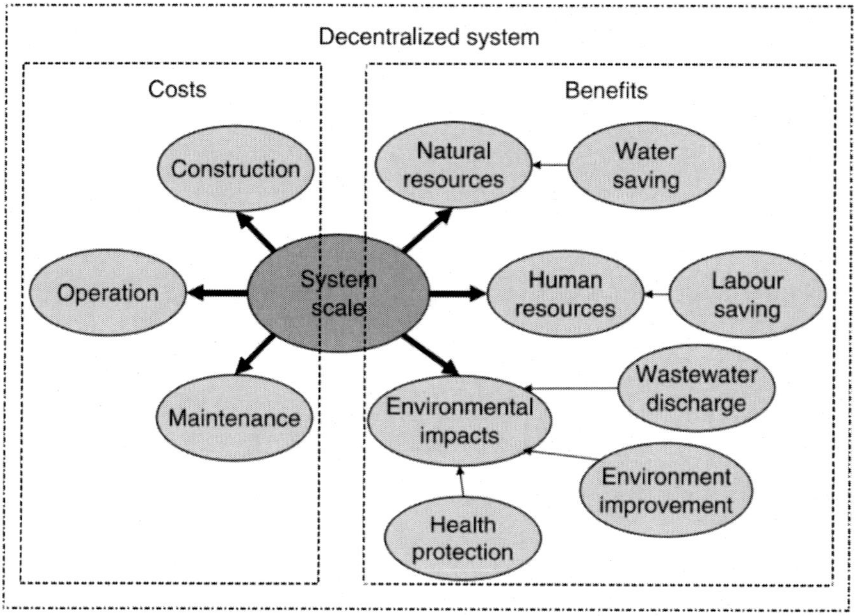

Figure 5. Relationship between factors related to costs and benefits (Chen and Wang, 2009).

The choice between a decentralized and centralized system must include a costs-benefits analysis. It should be noted that any particular benefit can be either positive or negative, especially when considering overall environmental impact (Chen and Wang, 2009; Walden et al., 2011). An example relationship between cost and benefit factors is shown in Figure 5.

Decentralized water management provides opportunities for better adaptation in the face of climate change and future uncertainties. Knowledge gaps regarding ongoing management requirements of a decentralized systems, as well as integration with existing centralized systems and potential implications (social, economic, environmental, etc.), need to be considered in a holistic way to identify potential opportunities and broader implementation of sustainable water management systems (Umar and Uhl, 2016).

WATER SENSITIVE URBAN DESIGN

Water managers have traditionally used pipes, ponds and other gray infrastructure designs to divert stormwater away from buildings, parking areas, and roadways. In contrast, proponents of Water Sensitive Urban Design (WSUD) attempts to managė stormwater near its origin, utilizing natural drainage pathways and best management practices (BMPs) (Tsegaye et al., 2019). WSUD is an urban design model that incorporates both natural and engineered systems to provide clean air, provide resilience to water scarcity, and offer a wide array of additional social and ecological benefits (Entrix, 2010; USEPA, 2016a). The concept relies heavily on indigenous parameters in order to create cost-effective, site-specific solutions to manage the concerns which preside over the area. A major focus of WSUD is the sustainable management of the three urban water streams: stormwater, wastewater, and potable water as shown in Figure 6. Although many see these three water streams as separate entities, they are intimately interrelated. WSUD capitalizes on this integration.

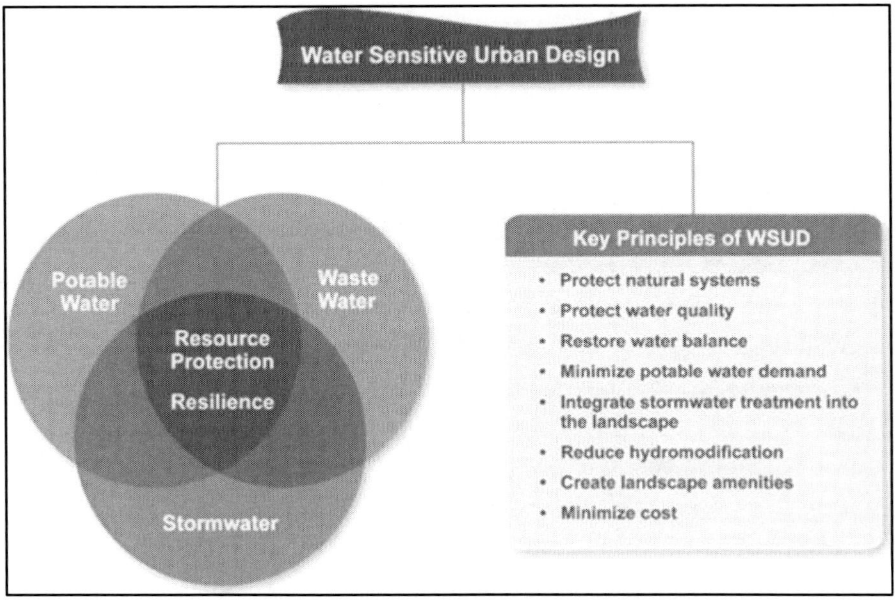

Figure 6. Key principles of water-sensitive urban design (Donofrio et al., 2009).

Stormwater is often regarded as an inconvenience and goes largely unrecognized for both its impacts and its potential. The rapid discharge of urban runoff in specific point locations contaminates freshwater sources, increases downstream flooding, increases municipal maintenance costs and disregards a perfectly suitable source of reclaimed water (Brinkmann & Goethe-Universitait, 1985; Burns et al., 2012; Jeng et al., 2005, Kim and Sansalone 2010). Contrary to conventional methods, WSUD attempts to manage stormwater at its source through BMPs that allow for the on-site or near-site treatment of runoff before it re-enters the watercourse (Kim and Sansalone 2008; Gallagher et al., 2018). Rain gardens, infiltration trenches, sedimentation basins, green roofs and vegetated swales are a few examples of aesthetically-functional stormwater BMP facilities that reduce the volume and velocity of runoff, attenuate pollutants, sequester nutrients, and offer additional aesthetic and social benefits to the local community (NCHRP, 2014; Entrix, 2010). These BMPs mainly rely on indigenous parameters including topography, vegetation and soil type to return cleaner water back to the watercourse via transpiration and/or percolation (Kabisch et al., 2016). Employing these natural processes permits the recharge of groundwater and protects surface water resources, thus replenishing and protecting local freshwater reserves. An incredible, yet lesser-known example of this was found in New York City, New York. The city was able to monetarily quantify the value of WSUD and ecosystems services within their water sector in the 1990s by purchasing pristine watershed land in the Catskill Mountains. Their initial $1.5 billion-dollar investment avoided nearly $8 billion in new stormwater treatment facilities that would have otherwise delivered the same filtration and purification required to protect their freshwater reserves (Benedict & McMahon, 2006).

Clever implementation of WSUD can also offer flood risk management at both local and regional scales (Sussams, 2015). Urban flooding often results when the capacity of the conventional municipal storm sewer system (MS4) is breached (Brown et al., 2008). As opposed to capturing and discharging acres of urban runoff through the conventional centralized system, treating runoff near its source will reduce volume inputs and alleviate stress on the MS4 (USEPA, 2014). On-site and near site attenuation of stormwater runoff will be 100% effective for small, intermittent runoff-causing events (Kim and Sansalone 2008). For the larger events, such a design storm, on-site and near-site attenuation will work in conjunction with the MS4 to ensure runoff is discharged efficiently.

One of the many successful examples of flood management is seen in the retrofits to the Lisbon Metropolitan Area in Portugal. The Ribeira de Sassoeiros was the discharge point for the surrounding municipality's runoff. Solutions to manage the increased flooding of the river were centered on WSUD through increasing localized infiltration. This approach successfully relieved a majority of the 18 cities within the municipal area from flood damage. Clever design of these infiltration facilities double as recreational parks and greenspaces for the local residents to enjoy during drier weather (Glendenning & Vervoort, 2015).

In locations where drought conditions are predicted to increase, the attenuation of heat is paramount. Some WSUD practices have the potential to significantly reduce a building's energy consumption. Trees and green roofs, for example, provide shade and reduce urban heat islands—factors which directly affect indoor heating and cooling costs (Sussams, 2015). Consequently, these features reduce emissions from power generation while simultaneously sequestering atmospheric carbon (Chen, 2015). Water loss is also abated when infiltration measures are incorporated. Green streets and green roofs provide a great example.

The city of Portland, Oregon is very progressive with WSUD retrofits. Many green facilities help to protect the community from stresses on local resources and climate impacts. The city's 2010 assessment of green facilities quantified their benefits. Green streets save an average of 155 kilowatt hours (kWh) per facility each year, and green roofs save an average of 8,270 kWh per acre each year. The vegetation inherent to the design of these facilities also offers the additional benefits outlined in Table 2. All of these benefits are additional to the water-sensitive management that also protects local water resources by curtailing nonpoint source pollution, treating runoff near its source and promoting groundwater recharge (Entrix, 2010).

Table 2. Benefits of Green Streets and Green Roofs (Entrix, 2010)

	Quantifiable Benefits			Social Effects			
	Energy Reduction	Air Quality Improvements	GHG Reductions	Human Well-Being	Aesthetic Improvements	Community Cohesion	Environmental Equity
Green Streets	155 kWh/facility/year	0.04 lbs/facility/year	0.3 metric tonnes/facility/year	Improvement to mental and physical health	3-5% increase in property values	Increases in social capital; decreased crime	Positive effect (depending on location)
Green Roofs	8,270 kWh/acre/year	7.7 lbs/acre/year	7.1 metric tonnes/facility/year	Improvement to mental and physical health	Possible positive effect	Possible increase in social capital	Positive effect (depending on location)

In addition to reductions in energy consumption, the quantifiable benefits of green streets and green roofs include air quality improvements and reductions in greenhouse gas emissions. Social benefits are also possible. Given the nature of green streets, social benefits are more pronounced; although, location plays a major factor. Green roofs offer a lesser extent of social benefits, largely due to their location.

Seoul is the capitol city of the Republic of Korea and maintains a population of more than 10 million (Seoul Metro Government, 2019). The major cause of flash flooding in the city is large, lengthy storm events (Kim, 2016). Green roof implementation proved to be a sustainable approach to mitigating these impacts. Analysis of these green roofs' attenuation capacity revealed the ability to capture between 10–60% of the total rainfall runoff in highly urbanized areas of Seoul (Shafique et al., 2018).

If we blatantly ignore the forewarnings of climatologists and continue on with business-as-usual, the resultant social, ecological and infrastructural damages could affect global gross domestic product (GDP) as much as 5.5%, with some studies estimating a loss of up to 20% (Harris et al., 2015; OECD, 2008). The loss of GDP will be felt differently between countries, and will greatly inflate issues of equity (ND-GAIN, 2014; Pielke et al., 2007). Fortunately, financial benefits also stem from the clever implementation of WSUD. An example of cost-reduction was reported by the City of Portland, Oregon. By indorsing the disconnection of gutter downspouts from storm sewers, installing vegetated swales along curb extensions and promoting infiltration practices for surface lot runoff, the city abated sewer system maintenance costs by roughly $0.89 to $4.08 per gallon of runoff treated on-site (Roseen et al., 2011). Additionally, The City of Normal, Illinois incorporated WSUD principles into a community redevelopment project. The $15.5 million renovation of a traffic circle was skillfully designed around aesthetically functional stormwater BMPs which later led to $160 million in private business investments, a 16% increase in nearby property values and a 46% increase in retail sales for the Uptown District (USEPA, 2016a).

Similar to social systems, ecosystems are anticipated to tolerate and adapt to some level of future climate change. The resiliency of ecosystems is essential for the protection of habitats, which are a major component of the services provided by successful WSUD implementation. WSUD has the ability to reverse habitat fragmentation and increase connectivity, cohesiveness, and overall ecosystem area (Sussams, 2015). Improving ecosystem connectivity, especially within urban areas, can aid in water resource protection and heat attenuation. Similarly, increased connectivity and coverage of ecosystems can provide certain buffers to extreme weather events.

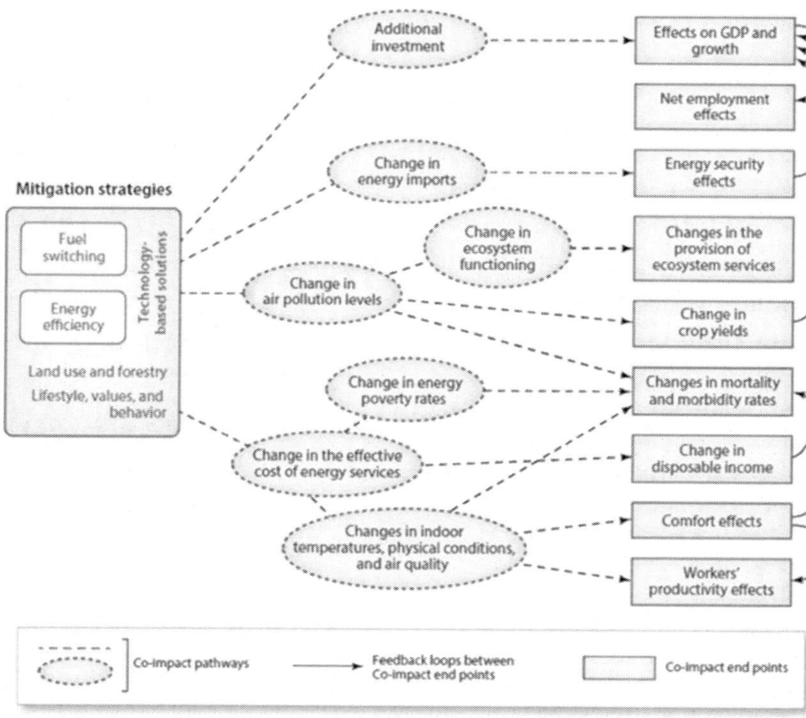

Figure 7. Co-benefits and welfare effects from climate change mitigation responses (Ürge-Vorsatz et al., 2014).

Recognizing the synergies and co-benefits of WSUD through mutually-supportive, cross-sector adaptations is particularly crucial within the urban environment. Ürge-Vorsatz et al., (2014) studied the interconnection of co-benefits and human welfare that result when employing sustainable energy practices to help mitigate climate impacts. They identified seven major categories of co-impacts from energy-based mitigation policies identified in the literature. Their conceptual framework is illustrated in Figure 7. By assessing different implementation strategies that promote alternate fuel sources and/or exercise energy efficiency, the impact of specific co-impact pathways on human welfare become apparent. Indirect influences between different welfare responses are also evident.

Intuitive WSUD design has the capability to satisfy many of the pathways in Figure 6. The case studies from New York City, Portland, and Seoul help to illustrate this. Following the topmost pathway, the additional investment these cities took in green facilities positively impacted GDP and growth though increases in surrounding property values, retail sales, and/or tax revenue.

Among the three bottommost pathways, change is evident in the effective cost of energy services, energy poverty rates, indoor temperatures, better physical conditions, and improved air quality. These, in turn, led to a shift in mortality rates via increased well-being, reallocation of disposable income through resource conservation, and increased comfort from improved well-being. Furthermore, indirect co-benefits are also evident. The feedback arrows at the far right of the diagram indicate that comfort effects are expected to reinforce both mortality rates and workers' productive success.

Application of GIS for Water Resource Management

The application of geographic information systems (GIS) for water resource management opens up incredible opportunities to help resource managers draw conclusions from predictive models and visualize spatially-correlated data.

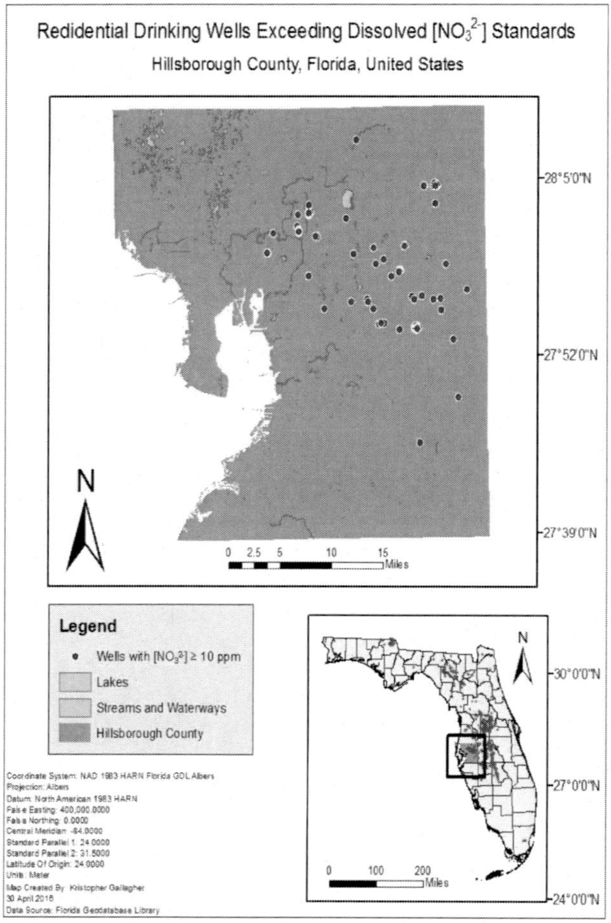

Figure 8. Contaminated residential drinking wells in Hillsborough County, Florida (USEPA, 2016b).

A clever use of GIS for water management is seen in Daegu, South Korea. Based on the current development of information and computer technology, a decision support system (DSS) was developed which aids in the identification and management of water resource problems. Using GIS Engine and SQL Server 2000 as web-based applications, the DSS provides managers with real-time forecasting of local hydrological and hydraulic conditions. It has helped to improve water managers' decision capabilities by optimizing decision-making by offering the ability to perform spatial queries and predictive scenario analyses. The successful application of the GIS and DSS in Daegu demonstrates the real-time efficiency and helps promote the move toward a fully-integrated urban water resource management decision support system for the South Korea (Zeng et al., 2012).

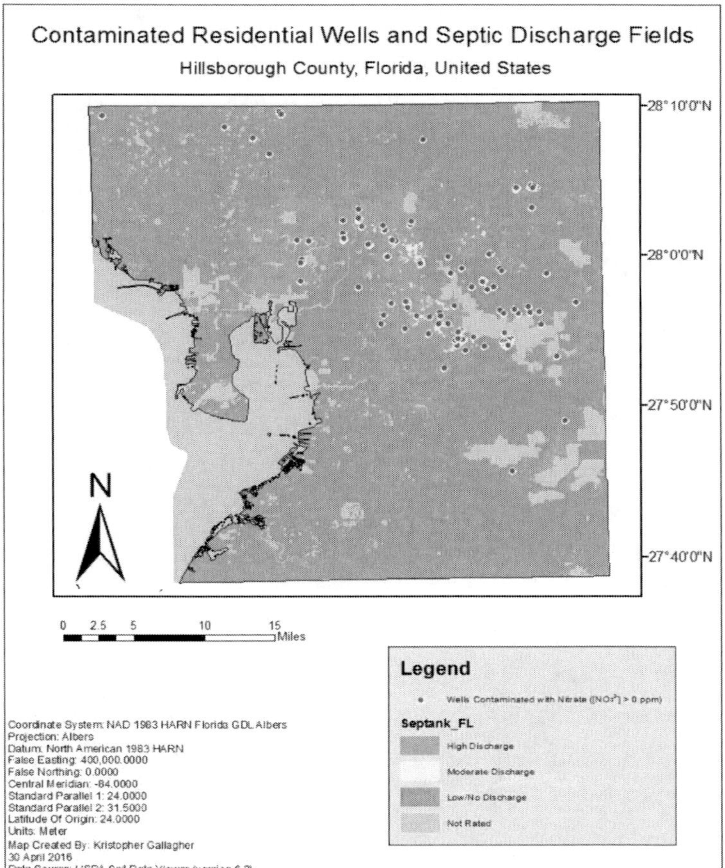

Figure 9. Contaminated residential drinking wells may be correlated to septic tank discharge volume (USEPA, 2016b).

As seen in the case study in Daegu, GIS is a powerful tool that can be harnessed to illustrate spatial concerns in real-time and help expedite effective decision-making. GIS allows users to easy visualize spatial patterns in a single dataset or across multiple datasets. Figure 8 depicts point data representing residential drinking wells where dissolved nitrates exceed the maximum contaminant level established by the EPA for drinking water standards. In this dataset, we see that most of the contaminated wells are clustered near the center of the county.

With the understanding that contamination originates from a source, we can assume that there must be a concentrated source of nitrates nearby the clustered points that is negatively impacting groundwater quality. Given that many of the central-county residents are serviced by drinking wells must indicate there is minimal water/wastewater servicing the area. Adding an additional data layer below the point data depicting septic tank discharge fields helps visualize how this potential source of nitrate contamination from septic tank discharge is correlated with the compromised drinking wells (Figure 9).

Additionally, a groundwater vulnerability model (GVM) could offer further insight into the impact of the nitrate plume on drinking wells in the area. A few well-known GVMs include DRASTIC [Depth-to-water, net Recharge, Aquifer media, Soils, Topography, Impact of vadose zone, hydraulic Conductivity model] (Yin et al., 2013), FAVA [Florida Aquifer Vulnerability Assessment] (FDEP, 2014), and MODFLOW (Harbaugh et al., 2000).

Just as a GVM could offer insight into the underground water nexus, a digital elevation model (DEM) can help assess the behavior of above-ground flow. In the water management discourse, DEMs are mainly used to assess the trajectories of runoff and floodwaters. Demand for precise and accurate DEMs in hydrography modeling is seen across the federal, state and local level. The United States Geological Survey (USGS) and the American Society for Photogrammetry and Remote Sensing (ASPRS) have established rigid base specifications that have become industry standard for high quality and accurate remotely-sensed geospatial data. Models produced under these standards, at minimum, will retain centimeter-grade accuracy (Heidemann, 2018). These base specifications were assembled for the launch of the national 3D Elevation Program (3DEP), which was implemented to systematically collect 3D elevation data for the contiguous United States,

Hawaii, Alaska and U.S. territories. With this program in place, precise DEMs are becoming readily and freely available to end-users.

Although DEMs are a key component in water resource management, additional data layers can help further refine results. As an example, GIS for water-sensitive urban design can be used to site stormwater infiltration BMPs across an area of interest. Growing interest in this field has yielded numerous GIS-based decision-support tools, such as the EPA's BMP Siting and the Grey to Green (G2G) Tool, for the sustainable management of stormwater runoff. The EPA's BMP Siting Tool and the G2G Tool aid end-users by locating potential suitable sites for the installation of a variety of different BMPs to manage stormwater runoff within a particular area of interest. These tools make use of three main types of BMPs—point, linear, and area—in order to address three main concerns with runoff: volume, peak discharge, and water quality (USEPA, 2014).

Before BMP siting can began, the appropriate data layers must be collected. The BMP Siting Tools requires the spatial data layers listed in Table 3. These different layers will be considered during the analysis, most of which will be used for the establishment of exclusion criteria where the BMP of interest cannot reside such as steep slopes, impervious surface area, private property, etc. Once the unsuitable sites are excluded, the remaining sites are analyzed based on peak BMP performance. For example, more permeable soils would be most suitable for infiltration BMPs.

Table 3. Data Layers: EPA's BMP Siting Tool

Input File	Layer Type
Digital Elevation Model	Raster
Land Use	Raster
Percent Impervious	Raster
Soil	Polygon Shapefile
Urban Land Use	Polygon Shapefile
Road	Polyline Shapefile
Stream	Polyline Shapefile
Groundwater Table Depth	Polygon Shapefile
Land Ownership	Polygon Shapefile

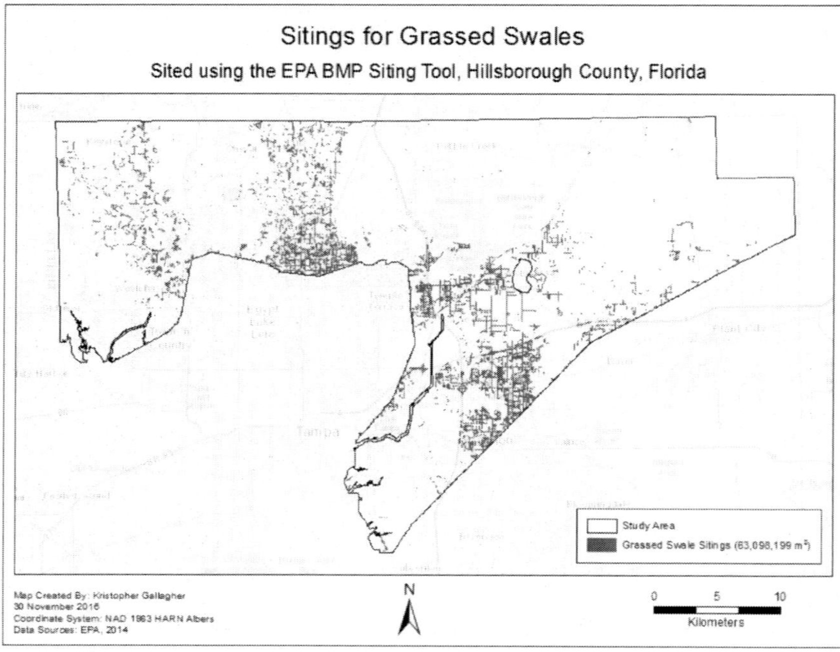

Figure 10. EPA BMP Siting Tool: Suitable bioswale sites.

Once the criteria have been established, the tool sites BMPs across the area of interest. Figure 10 illustrates the results of grassed swale sittings by the EPA's BMP Siting Tool throughout the northern portion of Hillsborough County, Florida. This study area covers about 951 km^2.

Conclusion

It has become increasingly obvious that the current practice of urban water management is not sustainable and the need to adopt innovative approaches is evident. Embracing sustainable water management approaches will pave the way for improved service and infrastructure. It includes efficiently providing safe, reliable, and easily accessible water to residents and businesses, offers reliable access to sanitation, and protects waterways from pollution. It also brings resilience and adaptability of water systems when faced with extreme weather events. Failure to address these challenges creates a myriad of potential health risks and often prevents underprivileged residents from contributing to the urban economy.

This chapter presented sustainable water management approaches which offer a diverse and versatile set of options for dealing with complex urban water challenges. These include: (1) a IUWM approach which offers interventions across the entire urban water cycle; (2) a decentralized water management approach that enhances the management of water and wastewater close to their sources, and (3) a water sensitive design (WSUD) management style that integrates the urban water cycle into the overall urban design to minimize environmental degradation while improving aesthetic and recreational appeal.

Critical to the success of these approaches is the early and continuous integration of all stakeholders in the planning and implementation process. The stakeholder participatory process is required to understand the priorities of potential users.

Despite the benefits, IUWM, WSUD and other decentralized approaches to water management have been slow to develop due to lack of knowledge, a reluctance to change existing practices and a lack of communication and partnerships. Barriers to implementation of these water management options will continue to exist until their effectiveness can be efficiently communicated by urban planners, environmentalist, urban foresters, landscape planners, engineers and other professionals. GIS can be used as an effective tool for storing, managing, and presenting spatial data often encountered in sustainable water management. The application of GIS as a smart mapping and communication tool could facilitate the uptake of technologies and the required paradigm shift presented in this book chapter.

REFERENCES

Ahmed, M. and Arora, M. (2012). Suitability of grey water recycling as decentralized alternative water supply option for integrated urban water management. *IOSR Journal of Engineering*, 2(9), 31-35.

Anderson, J. and Iyaduri, R. (2003). Integrated urban water planning: big picture planning is good for wallet and the environment. *Water Science and Technology*, 47 (7-8), 19-23.

Alcamo, J. M., Vörösmarty, C. J., Naiman, R. J., Lettenmaier, D. P. and Pahl-Wostl, C. (2008). A grand challenge for freshwater research: understanding the global water system. *Environmental Research Letters,* 3 1-6.

Athi Water Services Board (AWSB) (2012). Feasibility study and Master plan for Developing New Water Sources for Nairobi and Satellites Towns.

Benedict, M. A. and McMahon, E. T. (2006). Green infrastructure. Island, Washington, DC., 23-24. Retrieved 25 September 2016 from http://www.sprawlwatch.org/greeninfrastructure.pdf.

Bieker, S., Cornel, P. and Wagner, M. (2010). Semicentralised supply and treatment systems: integrated infrastructure solutions for fast growing urban areas. *Water Science and Technology*, 61(11), 2905-2913.

BMBF (2006). Semicentralized supply and disposal systems for fast growing urban regions in China, Final report of Sino-German Research Project. *German Federal Ministry of Science and Technology*.

Böhm, H. R., Schramm, S., Bieker, S., Zeig, C., Anh, T. H. and Thanh, N. C. (2011). The semicentralized approach to integrated water supply and treatment of solid waste and wastewater—a flexible infrastructure strategy for rapidly growing urban regions: the case of Hanoi/Vietnam. *Clean Technologies and Environmental Policy*, 13(4), 617-623.

Brinkmann, W. L. F. and J. W. Goethe-Universitait. (1985). Urban stormwater pollutants: Sources and loadings. *Geo Journa*l vol. 11, 277–283.

Brown, R., Keath, N. and Wong, T. (2008). *Proceedings from the 11th International Conference on Urban Drainage: Transitioning to water sensitive cities: Historical, current and future transition states.* Edinburgh, Scotland, UK.

Brown, R., Keath, N. and Wong, T. (2008). *Transitioning to Water Sensitive Cities: Historical, Current and Future Transition States.* 11th International Conference on Urban Drainage vol. 10, 1-10.

Burns, M. J., Fletcher, T. D., Walsh, C. J., Ladson, A. R. and Hatt, B. E. (2012). Hydrologic shortcomings of conventional urban stormwater management and opportunities for reform. *Landscape and Urban Planning* vol. 105, 230–240.

Chen, R. and Wang, X. (2009). Cost-benefit evaluation of a decentralized water system for wastewater reuse and environmental protection. *Water science and technology*, 59(8), 1515-1522.

Chen, W. Y. (2015). The role of urban green infrastructure in offsetting carbon emissions in 35 major Chinese cities: A nationwide estimate. *Cities* vol. 44, 112–120.

Choi, I. C., Shin, H. J., Nguyen, T. T., Tenhunen, J. (2017). Water Policy Reforms in South Korea: A Historical Review and Ongoing Challenges for Sustainable *Water Governance and Management. Water* 2017, 9, 717.

Cornel, P., Meda, A. and Bieker, S. (2011). Wastewater as a source of energy, nutrients and service water. *Treatise in Water Science*. Elsevier Publisher, 3, 92.

Cosgrove, W. J. and Loucks, D. P. (2015). Water management: Current and future challenges and research directions. *Water Resources Research,* 51, 4823– 4839. https://doi.org/10.1002/2014WR016869.

Daigger, G. T. (2009). Evolving urban water, and residuals management paradigms: Water reclamation and reuse, decentralization and resource recovery. *Water environment research*, 81(8), 809-823.

Donofrio, J., Kuhn, Y., McWalter, K. and Winsor, M. (2009). Water-sensitive urban design: An emerging model in sustainable design and comprehensive water-cycle management. *Environmental Practice* vol. 11, 179-189.

Domènech, L. (2011). *Rethinking water management: From centralised to decentralised water supply and sanitation models*. Documents d'anàlisi geogràfica, 57(2), 293-310.

Dopp, K. (2011). *Legislative Redistricting-Compactness and Population Density Fairness.* Available at SSRN 1945879.

Eckart, J., K. Ghebremichael, K. Khatri, S. Tsegaye and K. Vairavamoorthy (2012). "Integrated Urban Water Management for Nairobi." Report prepared for the World Bank by Patel College of Global Sustainability, University of South Florida, Tampa.

Entrix. (2010). Portland's Green Infrastructure: Quantifying the Health, Energy, and Community Livability Benefits, 2-1 thru 5-10. Retrieved 3 November 2016 from https://www.portlandoregon.gov/bes/article/298042.

FDEP [Florida Department of Environmental Protection]. (2014). Florida Aquifer Vulnerability Assessment. Retrieved 13 July 2016 from http://www.dep.state.fl.us/geology/programs/hydrogeology/fava.htm.

FGDL [Florida Geodatabase Library]. (2006). *Water supply restoration wells.* Retrieved 20 April 2016 from http://www.fgdl.org/metadataexplorer/explorer.jsp.

Gallagher, K. C., Alsharif, K., Tsegaye, S., and Van Beynen, P. (2018). A new approach for using GIS to link infiltration BMPs to Groundwater Pollution Risk. *Urban Water Journal*, 15(9), 847-857.

GAUFF (2011). *Arua Emergency Water Supply Project: Inception report.* Retrieved 21 July 2012 from http://www.amk.rwth-aachen.de/uploads/media/RECLAIM_WATERPublishable_Final_Activity_Report.pdf.

Gleick, P. H. (2002). *Dirty water: estimated deaths from water-related diseases 2000/2020.* Pac. Inst. Stud. Dev., Environ. Secur., Oakland, CA. Retrieved 3 September 2019 from http://www.pacinst.org/reports/water_related_deaths/water_related_deaths_report.pdf.

Gleick, P. H. (2009). Doing More with Less: Improving Water Use Efficiency Nationwide. *Southwest Hydrology*, 8(1), 20-21.

Glendenning, C. J. and Vervoort, R. W. (2015). Water as an Element of Urban Design: Drawing Lessons from Four European Case Studies. *Sustainable Water Use and Management* vol. 1, 17–43.

Harbaugh, A. W., Banta, E. R., Hill, M. C. and McDonald, M. G. (2000). MODFLOW-2000. The U. S. Geological Survey Modular Ground-Water Model-User Guide to Modularization Concepts and the Ground-Water Flow Process. Open-file Report. U. S. Geological Survey, (92), 134.

Harris, J. M., Roach, B. and Codur, A. (2015). The economics of global climate change. Global Development and Environment Institute. Tufts University, 19-25.

Heidemann, H. K. (2018). Lidar base specification (ver. 1.3, February 2018): *U.S. Geological Survey Techniques and Methods*, book 11, chap. B4, 101 p., Retrieved 3 September 2019 from https://doi.org/10.3133/tm11b4.

Herrera, M., Canu, S., Karatzoglou, A., Perez-García, R. and Izquierdo, J. (2010). An approach to water supply clusters by semi-supervised learning. Proceedings of iEMSs.

Hill, C. (2016). Water Sustainability and How to Achieve It. *Water Finance and Management.* December 7, 2006. Retrieved https://waterfm.com/water-sustainability-achieve/.

Huang, D., Vairavamoorthy, K. and Tsegaye, S. (2010). Flexible design of urban water distribution networks. In *World Environmental and Water Resources Congress 2010: Challenges of Change* (pp. 4225-4236).

ICLEI (2011). SWITCH Training kit module 1.

IMF (2019). International Monetary Fund World Economic Outlook (October - 2018), International Monetary Fund (IMF),Washington, D.C. Retrieved 1 August 2019 from https://www.imf.org/en/data.

Jacobsen, M., Webster, M. and Vairavamoorthy, K. (Eds.) (2012). The Future of Water in African Cities: Why Waste Water? Washington, D.C.: The World Bank.

Jeng, H. A. C., Englande, A. J., Bakeer, R. M. and Bradford, H. B. (2005). Impact of urban stormwater runoff on estuarine environmental quality. *Estuarine, Coastal and Shelf Science* vol. 63, 513–526.

Kabisch, N., Frantzeskaki, N., Pauleit, S., Naumann, S., Davis, M., Artmann, M. and Zaunberger, K. (2016). Nature-based solutions to climate change mitigation and adaptation in urban areas: perspectives on indicators, knowledge gaps, barriers, and opportunities for action. *Ecology and Society* vol. 21, 1-15.

Kim, J. Y. and Sansalone, J. J. (2008). Hydrodynamic Separation of Particulate Matter Transported by Source Area Runoff. *Journal of Environmental Engineering*, 134(11), 912-922.

Kim, J. Y. and Sansalone, J. J. (2010). Representation of Particulate Matter COD (Chemical Oxygen Demand) in Urban Source Area Rainfall-Runoff. *Water Air and Soil Pollution*, 205, 113-132.

Kim, R. (2016). Low Impact Development and Green Infrastructure in South Korea: Trends and Future Directions. *Ecol. Resil. Infrastruct.*, 3, 80–91.

Leas, E. C., Dare, A. and Al-Delaimy, W. K. (2014). Is gray water the key to unlocking water for resource-poor areas of the Middle East, North Africa, and other arid regions of the world?. *Ambio,* 43(6), 707–717. Retrieved 10 September 2019 from https://link.springer.com/article/10.1007%2Fs13280-013-0462-y.

Long, J. (2001). Desecuritizing the Water Issue in Singapore—Malaysia Relations. Contemporary Southeast Asia, 23(3), 504-532. Retrieved 5 September 2019 from http://www.jstor.org/stable/25798564.

Mitchell, V. G. (2004). Integrated Urban water management: A rview of current Australian practice. Canberra: Australia. Australiam Water Conservation and Reuse Research Program.

Nam, N. H. (2006). Decentralized wastewater treatment and Reclamation using membrane bioreactor. Asian Institute of Technology.

NCHRP [National Cooperative Highway Research Program]. (2014). National Research Council (U.S.)., American Association of State Highway and Transportation Officials., Oregon State University., University of Florida., GeoSyntec Consultants., and Low Impact Development Center, Inc. (2006). Evaluation of best management practices for highway runoff control. Washington, D.C: Transportation Research Board, 40-79.

ND-GAIN Country Index. (2014). ND-GAIN. Retrieved 21 September 2016 from http://index.gain.org/.

Newman, P. (2001). Sustainable urban water systems in rich and poor cities--steps towards a new approach. *Water science and technology: a journal of the International Association on Water Pollution Research*, 43(4), 93.

OECD [Organisation for Economic Co-operation and Development]. (2008). *Climate change mitigation: What can we do?,* 11-18. Retrieved 30 September 2016 from www.oecd.org/env/cc/41751042.pdf.

Parkinson, J. N., Goldenfum, J. A. and Tucci, C. (2010). Integrated Urban Water Management: Humid Tropics: UNESCO-IHP. CRC Press.

Pielke, R., Prins, G., Rayner, S. and Sarewitz, D. (2007). Climate change 2007: Lifting the taboo on adaptation. *Nature* vol. 445, 597-598.

Roseen, R. M., Janeski, T. V., Houle, J. J., Simpson, M. and Gunderson, J. (2011). Forging the link: Linking the economic benefits of low impact development and community decision. The UNH Stormwater Center, Univ. of New Hampshire, Durham, NH, 3-20 thru 3-26.

Seoul Metropolitan Government (2019). *Historical Seoul Population.* Retrieved 25 June 2019 from http://english.seoul.go.kr/get-to-know-us/seoul-views/meaning-of-seoul/4-population.

Shafique, M.; Kim, R.; Kyung-Ho, K. (2018). *Green Roof for Stormwater Management in a Highly Urbanized Area: The Case of Seoul*, Korea. Sustainability, vol. 10, 584.

Smaniotto Costa, C., Norton, C., Domene, E., Hoyer, J., Marull, J. and Salminen, O. (2015). Water as an Element of Urban Design: Drawing Lessons from Four European Case Studies. In: Leal Filho W., Sümer V. (eds) *Sustainable Water Use and Management. Green Energy and Technology*. Springer, Cham.

Sussams, L. W., Sheate, W. R. and Eales, R. P. (2015). Green infrastructure as a climate change adaptation policy intervention: Muddying the waters or clearing a path to a more secure future? *Journal of Environmental Management* vol. 147, 184–193.

Tsegaye, S., Eckart, J. and Vairavamoorthy, K. (2012). Urban Water Management in Cities of the Future: Emerging Areas in Developing Countries. *On the Water Front*, 42.

Tsegaye, S., Singleton, T. L., Koeser, A. K., Lamb, D. S., Landry, S. M., Lu, S. and Ghebremichael, K. (2019). Transitioning from gray to green (G2G)—A green infrastructure planning tool for the urban forest. *Urban Forestry and Urban Greening*, 40, 323-334.

Tsegaye, S. and Vairavamoorthy, K. (2009). Agent-Based Modeling to Estimate Residential Water Demand and to Explore Optimal Demand Side Water Management strategies. WEDC: Water Engineering

Development Centre, SWITCH: Managing Water for the City of the Future.

Umar, M. and Uhl, W. (2016). Integrative review of decentralized and local water management concepts as part of smart cities (lowasmart).

Ürge-Vorsatz, D., Herrero, S. T., Dubash, N. K. and Lecocq, F. (2014). Measuring the co-benefits of climate change mitigation. *Annual Review of Environment and Resources* vol. 39, 549-582.

USEPA. (2014). *Coastal Stormwater Management through Green Infrastructure: A Handbook for Municipalities*, 50-65. Retrieved 25 October 2016 from http://www.mass.gov/eea/docs/mbp/publications/massbays-green-infrastructure-handbook.pdf.

USEPA. (2016a). *City green: Innovative green infrastructure solutions for downtowns and infill locations*, 1-75. Retrieved 3 October 2016 from www.epa.gov/smartgrowth.

USEPA. (2016b). Table of regulated drinking water contaminants. Retrieved 21 April 2016 from https://www.epa.gov/your-drinking-water/table-regulated-drinking-water-contaminants#one.

Vairavamoorthy, K. (2011). Proceedings from the NARST Annual International Conference: New urban leaders for sustainable cities of the future. Orlando, FL.

Vairavamoorthy, K., Gorantiwar, S.D. and Mohan, S. (2007). Intermittent water supply under water scarcity situations *Water International*, 32(1), 121 -132.

Van der Steen, P. and Howe, C. (2009). Managing water in the city of the future; strategic planning and science. *Reviews in Environmental Science and Bio/Technology*, 8(2), 115-120.

Verstraete, W. and Vlaeminck, S. E. (2011). ZeroWasteWater: short-cycling of wastewater resources for sustainable cities of the future. *International Journal of Sustainable Development and World Ecology*, 18(3), 253-264.

Walden, A., Nahm, M., Barnett, M. E., Conde, J. G., Dent, A., Fadiel, A. and Eisenstein, E. L. (2011). Economic analysis of centralized vs. decentralized electronic data capture in multi-center clinical studies. *Studies in health technology and informatics*, 164, 82–88.

Wang, X., Chen, R., Zhang, Q. and Li, K. (2008). Optimized plan of centralized and decentralized wastewater reuse systems for housing development in the urban area of Xi'an, China. *Water science and technology*, 58(5), 969.

Water World (2012). A Shining Star in Rainwater Harvesting, Water and Wastewater International, September 2012. Retrieved 30 June 2019 from https://www.waterworld.com/international/desalination/article/16201474/south-korea.

Weber, B., Cornel, P. and Wagner, M. (2007). *Semi-centralized supply and treatment systems for (fast growing) urban areas*. 55(1-2), 349-356.

World Water Council (2015). S. Korea showcases smart water management, 7th World Water Forum Daegu, South Korea. April 12-17, 2015. Retrieved 30 June 2019 from https://en.yna.co.kr/view/AEN20150416009700320#none.

Yin, L., Zhang, E., Wang, X., Wenninger, J., Dong, J., Guo, L. and Huang, J. (2013). A GIS-based DRASTIC model for assessing groundwater vulnerability in the Ordos Plateau, China. *Environmental Earth Sciences,* 69(1), 171-185.

Zeng, Y., Cai, Y., Jia, P. and Jee, H. (2012). Development of a web-based decision support system for supporting integrated water resources management in Daegu city, South Korea. *Expert Systems with Applications,* 39, 10091-10102.

Zhou, Y., Vairavamoorthy, K. and Mansoor. MAM (2009). Integration of urban water services, *Desalination*, 248:402-409.

In: Sustainable Water
Editors: V. Gnaneswar Gude et al.
ISBN: 978-1-53617-335-2

Chapter 2

WATER RESOURCES MODELING: MODEL SELECTION, VALIDATION AND UNCERTAINTY ANALYSIS

Mengistu Geza* and Ali Shojaeizadeh
Department of Civil and Environmental Engineering,
South Dakota School of Mines and Technology,
Rapid City, South Dakota, US

ABSTRACT

Watershed models can be used to help determine potential sources of pollutant loading in natural water systems. Planners and regulators use watershed models to evaluate the impact of various scenarios including climate change, urbanization and other land use decisions. Models have also been used to evaluate groundwater contamination problems that evolve from hazardous waste sites or landfills. A range of models are available with varying degree of complexity. Complex models require expertise to use and relatively more data to build. Simpler models are less costly to use, but yield results that are much more uncertain, and are also

---

* Corresponding Author's Email: stu.geza@sdsmt.edu.

less defensible because they generally do not represent actual conditions with respect to hydrology or pollutant fate and transport. Thus, model selection and assessment of the validity of a watershed model is a challenging task. This chapter attempts to provide guidance on how to choose and apply models of varying complexity. The chapter presents different types of models, model selection process, calibration and uncertainty analysis. A guidance is provided on selection of appropriate model for a particular decision. The chapter includes methods for model calibration and risk-based decision-making. There is uncertainty associated with the input data used in the model, and with the model itself that result in uncertainty in the model's output data. This uncertainty should be considered in the decision-making process. Methodologies for model uncertainty analyses are presented.

Keywords: watershed modeling, model selection, calibration, uncertainty

INTRODUCTION

The chapter provides an overview of model types, model selection, calibration and uncertainty analysis with emphasis on watershed-scale modeling. It provides guidance about the type of model that should be used for particular scenarios. Planners and regulators, and other decision-makers apply watershed-scale models to assess the impacts of natural and anthropogenic effects such as climate change, urbanization, and other land use decisions on the environment. The U.S. EPA is promoting a watershed-scale approach with respect to permitting, mitigating, apportioning, or evaluating pollutant and nutrient loading from various sources. Thus, watershed-scale models of varying complexity are important tools for making quantitative assessments. Models, when applied properly, provide useful information for decision-making. Models can be very useful to evaluate the relative impact of different land uses on surface water and groundwater quality at local and watershed-scales. Watershed-scale models have been increasingly applied to agricultural problems in the last decade. Mathematical models are frequently used to evaluate groundwater contamination problems at scales of a few square kilometers or less, usually for groundwater plumes that evolve from hazardous waste sites or landfills.

Watershed models have been developed ranging from simple screening type models to more complex models. Complex models can be expensive to use, partially because of the labor cost for experts who can apply the models, but also because of the cost of acquiring the appropriate data to build reliable models. As a general rule, simpler models are less costly, but yield results that are much more uncertain, and are also less defendable because they generally do not represent actual conditions with respect to hydrology or pollutant fate and transport. Thus, screening models and tools that implement simplifying and conservative assumptions should be used first to evaluate if impacts due to anthropogenic activity are a potential concern within the limits of uncertainty associated with the model results. If impacts are shown to be a reasonable concern, then models with increasing complexity and implementation cost can be used in a sequential manner until the decision-makers are comfortable with the model predictions, including the model uncertainty, relative to the risk associated with the decisions. This chapter attempts to demonstrate the potential benefits of using mathematical models and provides guidance on how to choose and apply models of varying complexity.

Chapter 2 explains the different types of models, the varying complexity of model classes, and how they represent hydrological systems. Chapter 3 presents a guide to model selection. Chapter 4 presents model calibration, validation and uncertainty analysis.

Model Types

Different Types of Equations in the Hydrologic System

This section focuses on classification of models based on complexity of approaches/equations used to represent models ranging from simple empirical equations to numerical solutions for differential equations.

Mass Balance Equations

Any hydrologic system can be represented by applying the law of conservation of mass. When constructing a mass balance model the mass (e.g., of water or nutrients) entering a system must either leave the system or accumulate within the system. Mass balance equations take the form of:

$$Input - Output = Accumulation$$

The boundaries of the system (e.g., watershed) must be well-defined so that inputs and outputs crossing the system boundary can be quantified or estimated. Mass balance equations can be applied to hydrologic systems by using measured data of inputs/outputs, approximating unmeasured parameters using empirical or analytical equations, and making simplifying assumptions. Common simplifying assumptions may include: 1) neglecting quantities with magnitudes considerably less than the other quantities in the mass balance equation, and 2) steady-state condition, which makes the accumulation term zero.

Empirical Equations

An empirical equation is one based solely on observation rather than theory and forms a relationship between a desired, difficult to measure parameter and a commonly measured parameter. An empirical relationship requires only confirmatory data irrespective of theoretical basis. One of the cases where empirical equation was applied is the use of the empirically based curve number (CN) method for rainfall-runoff modeling (USDA Soil Conservation Service, 1972). The method has been widely used because of its simplicity and data availability.

The Holtan's equation for infiltration is an empirical equation based on a storage concept. The equation was developed at USDA, Agricultural Research Service to estimate infiltration using information that was generally available or could be readily obtained for major soils of the nation (Holtan, 1967). The Horton's equation is another viable option for estimating infiltration rates. The Horton's equation is an empirical infiltration equation derived from the simple assumption that infiltration declines as rain continues (Horton, 1939).

Analytical Equations

Analytical solutions are possible for mathematical models that have a closed-form solution. In other words, changes in a system can be expressed as a mathematical function with one exact solution. These solutions often assume a steady-state or only apply to models with a single time-step. For example, to calculate infiltration rate, the physically-based models describe the infiltration process in detail, and Richards model (Richards, 1931) and the Green-Ampt model (Green and Ampt, 1911) are those most commonly used for simulating the infiltration process. The solution to Richard model requires an iterative implicit numerical technique with fine discretization in time and space; thus, this model is computationally intensive. The Green-Ampt model is an analytical solution derived from Darcy*'s* equation and is derived indirectly through Richards model with some simplification.

Numerical Equations

Numerical solutions are required when the governing equations are ordinary differential equations (ODE) or partial differential equations (PDE). Numerical solutions do not seek exact answers because exact answers are impossible to obtain in practice. Numerical approximations are achieved using a computer code that solves a set of algebraic equations that approximate the governing equation, boundary conditions, and initial conditions of the mathematical model. These solutions are often necessary when multiple time-steps are required to simulate processes in the watershed. The approximate solution to one ODE or PDE becomes an input to another ODE or PDE in a time-series analysis. An example of a numerical equation based on a nonlinear partial differential equation is Richards' infiltration equation that represents the movement of water in unsaturated soils (Richards, 1931).

Categories of Hydrologic Models

Hydrologic models can be grouped into several categories based on physical domain/compartment (groundwater model, surface water model or integrated), spatial representation (lumped or distributed), temporal representation (single or continuous simulation), method of solution (analytical or numerical), level of complexity (screening level or well detailed/complex) and spatial compartment (surface water, groundwater or integrated models). A screening model is any model used to evaluate a system using highly simplified relationships between system inputs and outputs and requiring minimal data. When completing watershed-scale modeling screening models can be used to identify potentially sensitive areas, to highlight areas where more data are needed, or to quickly test alternative conceptual models. This type of model can be used to determine whether more resources should be devoted to data gathering or toward the development of more complex models or quantitative tools. Consequently, screening models provide cost-effective evaluations on strategies before more expensive physically-based models are employed for a specific site. Increased complexity normally translates to higher costs for the gathering of necessary data, implementation, and calibration of the model.

An overview of only a few of the categories is provided. Hydrologic models classifications based on the level of complexity (screening versus complex) and on physical domain (groundwater, surface water, and integrated) are described here.

Mass Balance Screening Models

Mass balance screening models are usually developed in a spreadsheet and are highly simplified estimates of water and nutrient balances. Mass balance screening models produce a conservative estimate of nutrient loads, which helps the user to 1) ensure that there is no major problem in a given site and no further modeling is required, or 2) determine that there is a potential risk for flooding or pollution and that additional assessment is needed which eventually lead to more complex model types. For example, a mass balance model, On-Site and Fertilizer (OSF) model was applied for land management purposes in Martha's Vineyard, Massachusetts (Gaines, 1986) and is the simplest screening model. OSF assumes that N inputs in the watershed are from homes, and simply adds an annual load of septic N (6.8 kg N/year) and lawn fertilizer N (4.8 kg N/year) per household. Atmospheric N loads are assumed to be taken up within the watershed (Valiela et al., 2002). The number of homes in the watershed is the only data requirement, so it will provide a general estimate of N loading to the mouth of the watershed compared to more complex models of nitrogen transport. If a more detailed analysis is required, complex numerical models that include detailed processes such as advection, dispersion, and reaction rates can be implemented.

GIS Based Screening Models

GIS based screening models are implemented either completely within a GIS or require the use of GIS for processing of geographically-referenced data (e.g., land use, soil property, and precipitation) as primary model inputs. A GIS based screening model is a model that combines a simple conceptual model and analytical equations that can be solved using a GIS. It generally captures the regional spatial characteristics of a system without including the detailed processes involved or data density that would be necessary for site-specific delineations. When integrated within GIS, computations become powerful and time-efficient because of advantages in data storage, manipulation, analysis, and visualization of spatial data. This may include a simplified vulnerability assessment without going into details of numerical modeling. For example, Nitrogen Loading Model (NLM) predicts total dissolved N loads to shallow estuaries from rural suburban watersheds where groundwater flow is the dominant transport vehicle. The NLM is complex enough to represent the nature of the systems and sufficiently adapted to local conditions to produce accurate predictions, while simple enough to be applied to different types of situations. NLM uses values for per capita contributions to estimate N inputs from DWTS that are derived from GIS data representing the total number of residences. NLM also accounts for losses of nitrogen within septic systems (by denitrification, volatilization of ammonia, or by adsorption of ammonium) as well as within leaching fields (Valiela et al., 1997; Valiela et al., 2002).

Surface Water Models

Surface water models focus on fluxes between the atmosphere and ground-surface and changes in storage within rivers, streams and other surface water bodies. A surface water model normally combines empirical models and/or numerical models to simulate flow rates and nutrient concentrations over various time-scales. HEC-RAS is an example of a river hydraulics model developed by the U.S. Army Corps of Engineers to compute one-dimensional water surface profiles for steady or unsteady flow (Hydrologic Engineering Center, 1995, HEC-RAS River Analysis System). River2D model, a two-dimensional, depth-averaged hydrodynamic and fish habitat model developed specifically for use in natural streams and rivers. A complete description of the formulation and implementation of the model is contained in Ghanem et al. (1995).

Subsurface Models

Vadose Zone Models

Vadose zone models are built around the unsaturated zone and consider fluxes between the overlying ground surface and the underlying saturated zone. For example, HYDRUS numerically solves the Richards equation for saturated-unsaturated water flow and convection-dispersion equations for heat and solute transport (Simunek et al., 1999). The transport equations include provisions for reactions between the solid and liquid phases along with first-order degradation reactions.

Groundwater Models

Groundwater models focus on changes in storage and fluxes within the saturated zone. Normally a combination of empirical models and/or numerical models that can predict hydraulic heads and nutrient concentrations over various time-scales. For example, the MODular three-dimensional groundwater FLOW (MODFLOW) model numerically solves the groundwater flow equation for the saturated porous medium by using a finite-difference method (Hill, 1992). MODFLOW can be used in conjunction with compatible modules such as MT3D (Zheng and Wang, 1999; Zheng, 1993) or RT3D (Clement 1997, Clement et al., 1998) to simulate contaminant transport and reaction. It can also be used in conjunction with other models to allow for modeling of flow and transport coupled between surface water, vadose zone, and saturated groundwater zones.

Integrated Watershed Models

Integrated watershed models include surface water, vadose zone, and groundwater flow modules; thus, allowing for temporal calculations of flux and storage in each component of the hydrologic system. Integrated hydrologic models that include land surface modeling, unsaturated zone, and groundwater flow provide a more sophisticated physical description of water fluxes across compartment boundaries. ParFlow-CLM includes snowmelt process, unsaturated zone flow, water table fluctuation, and groundwater flow.

Thus, remote sensing products such as soil moisture, surface temperature, and snow cover or leaf area index could be assimilated to improve the accuracy of predictions (Maxwell and Miller, 2005). For example, the Soil and Water Assessment Tool (SWAT) model is a watershed model most often used in Total Maximum Daily Load (TMDL) assessments. SWAT is known for its ability to simulate many hydrologic processes such as surface runoff, lateral flow, groundwater flow, sediment loading, N, P, and pesticides. SWAT operates on a daily time step and requires, at a minimum, the following input data: climatic, digital elevation, land use-land cover, and soils data.

SWAT was developed to predict the impact of land management practices on water, sediment, and agricultural chemical yields in large complex watersheds that have varying soils and land use and management practices over a long period of time (Borah et al., 2006). Other models in this category include Hydrological Simulation Program – Fortran (HSPF, Donigan, et al., 1984), Systeme Hydrologique European (MIKE-SHE,

Danish Hydraulic Institute (DHI), 1998), Watershed Analysis Risk Management Framework (WARMF, Herr et al., 2001).

Model Selection Process

Introduction

Watershed-scale models can be useful tools for assessing anthropogenic impacts, as well as the relative impacts of one factor compared to the impacts of another factor. In general terms, because the hydrology and chemical transport processes at the watershed-scale are so complex, numerical models are necessary to provide a rigorous assessment. Simple models can be used as screening tools, or for gaining a basic understanding of how changes in watershed inputs influence watershed-scale water quality.

Selection of the appropriate model depends on the intended use of the model and the desired output. The ideal model for quantitative watershed-scale assessments would have the following attributes.

1. Able to simulate both water quality and water quantity issues.
2. Able to simulate pollutants relevant to watershed management.
3. Capable of simulating all relevant compartments of the hydrologic system
4. Include the relevant chemical reactions in soil, groundwater and surface water.
5. Able to account for additions and losses of contaminants via the most important mechanisms.

Finally, the model selection will always be linked closely to the question at hand including the decision(s) being made and the importance of making correct decisions. For example, nearly all models that are good at handling surface water processes have simplistic representations of groundwater processes.

The converse is true for models that excel at simulating groundwater processes. Even though recent advancements have been made to integrate and simulate the physical interactions between surface and groundwater, very few models can handle contaminant transport and reactions associated with the surface water and groundwater interactions. Thus, the final model selected may differ depending on the contaminant (e.g., viruses versus nitrogen versus pharmaceuticals) because not all models can handle all chemical transport and reaction processes.

Ultimately, a model recommended for general use should be a widely used model that is available in the public domain. Therefore, only those models that have been vetted by significant prior use will be accepted by technical and regulatory communities and will be able to withstand legal or technical challenges.

Finally, it is important to emphasize that the type and capability of the model selected should be consistent with the decisions that the modeling is intended to support.

Factors to Consider in Model Selection

When selecting the type of model that is necessary for watershed-scale assessment, the primary factors to consider are 1) status of watershed, 2) regulatory and environmental issues, 3) hydrologic, chemical, and physical processes, 4) relative costs of a model, 5) information needs, and 6) pollutant fate and transport processes. These factors can provide guidance on the selection of the appropriate model type (Figure 1).

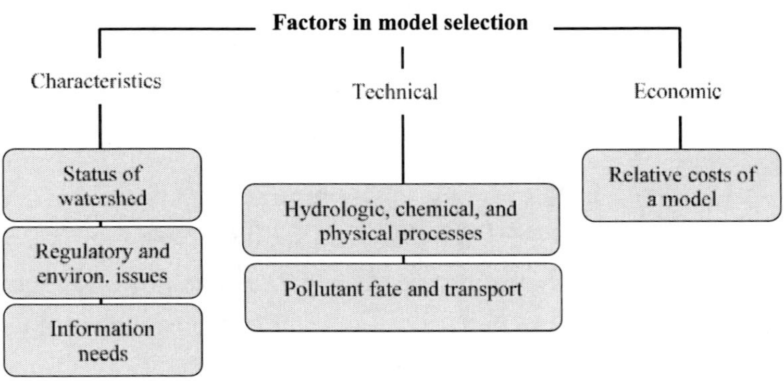

Figure 1. Factors to consider in model selection.

Status of Watershed

An evaluation of the status of the watershed in terms of pollution trends (impaired versus healthy watershed) should be completed to understand the relationship between current water quality and established criteria. Increasing concentrations (worsening) of nutrients or other indicators would suggest a need for a more complex model. In these cases, a more rigorous model could be used to model water quality conditions under current land use and wastewater disposal practices, while predicting the outcomes of alternative practices.

Regulatory and Environmental Issues

If the surface or groundwater at risk from contamination is also a source of drinking water in the watershed, there may a significant human health risk. Resources and thus consequences are not generally independent. Thus, if one resource is impacted it may lead to several consequences.

For example, contamination of a shallow aquifer used for drinking water might result in regulatory action requiring advanced pre-treatment from agriculture, industry or a developer may be necessary before pollutant are discharged into streams. For instance, if the shallow aquifer discharges nutrients to streams via base flow, aquatic habitat might be degraded, aesthetic and recreational values might be lost, and economic losses from tourism might occur.

Hydrologic, Chemical, and Physical Processes

In areas where groundwater is typically an important source of drinking water, individuals obtain water supply from individual wells, and the source could usually be a shallow aquifer that would be vulnerable to contamination from surface sources. If there is a community water system, wells may tap deeper aquifers that are less likely to be impacted from surface sources. However, if the source of drinking water is a shallow (water table) aquifer, evaluation of alternatives for protecting the aquifer would likely require a model capable of simulating nutrient movement from the land surface into the subsurface may be needed (e.g., unsaturated zone models). An integrated saturated and unsaturated zone model with the capability to model flow and chemical transport in both compartments may be needed if the groundwater is located downgradient of a pollutant source.

In areas where surface water is a primary source of drinking water, or where aquatic habitat is a resource at risk from contamination from surface sources, models capable of simulating surface water and soil processes might be most appropriate. Thus, understanding pathways for chemical transport is needed to select an appropriate model.

Relative Costs of a Model

Financial resources may limit the choice of model type to a more simplistic and inexpensive model. However, if for example, human or animal health is at risk, a complex (and thus generally expensive) model is necessary regardless of cost. In such cases, the damage costs cannot even be

Table 1. Modeling cost, complexity, and level of information generated

Model Type	Cost components and level of complexity				
	Training Cost	Operation Cost	Cost of Data	Level of Complexity	Information Generated
SWMM	Low	Low	Low	Low	Low
OSF	Low	Low	Low	Low	Low
PLS	Low	Low	Low	Low	Low
NLM	Low	Low	Low	Low	Low
MANAGE	L-A	L-A	L-A	L-A	L-A
GWLF	L-A	L-A	L-A	L-A	L-A
GLEAMS	Average	Average	Average	Average	Average
ANSWERS	Average	Average	Average	Average	Average
CREAMS	Average	Average	Average	Average	Average
AGNPS	A-H	A-H	A-H	A-H	A-H
WARMF	High	High	High	High	High
SWAT	High	High	High	High	High
HSPF	High	High	High	High	High
MIKE-SHE	High	High	High	High	High

L-A = Low to Average, A-H = Average to High

SWMM = Storm Water Management Model, OSF = On-Site and Fertilizer model, PLS = Pollutant Load Screening model, NLM = Nitrogen Loading Model, MANAGE = Method for Assessment, Nutrient-loading, and Geographic Evaluation of Nonpoint Pollution model, GWLF = Generalized Watershed Loading Function model, GLEAMS = Groundwater Loading Effects of Agricultural Management Systems, ANSWERS = Areal Nonpoint Source Watershed Environment Response Simulation, CREAMS = Chemicals, Runoff, and Erosion from Agricultural Management Systems, AGNPS = Agricultural Non-Point Source Pollution model, WARMF = Watershed Analysis Risk Management Framework, SWAT = Soil & Water Assessment Tool, HSPF = Hydrological Simulation Program – FORTRAN, MIKE-SHE = Systeme Hydrologique European model estimated in monetary terms. Users or regulators should use more complex models to evaluate possible abatement scenarios, which would be still far cheaper than the damage costs.

The prediction accuracy of models depends on how well model input spatial parameters describe the characteristics of the watershed (Geza et al., 2008). Thus, when there is high environmental risk, more detailed, high resolution data should be used. Under other conditions, simpler models could be used to provide a conservative estimate. The relative cost of the model is directly related to the time involved in gathering necessary data, implementation and calibration of the model.

Subjective assessment of watershed models that includes costs required for training a model user, model operation (model setup and calibration) and cost of data collection, level of model complexity and corresponding level of information generated is shown in Table. Models could be relatively inexpensive and less complex generating a limited amount of information or could be relatively expensive and more complex generating more information.

The best alternative model would be a simple model that costs less in terms of all cost components (set up, calibration, training, etc.) and at the same time is able to generate sufficient and accurate information.

Information Needs

In some watersheds, data may be limiting the use of a complex model. For example, data for a site may have shortcoming on geographic coverage, on continuity of periodic measurements, and on consistency regarding how samples were collected, analyzed and reported. Even when other criteria indicate that a more complex model should be used for a site, users may opt for a simple model formulation with fewer data requirements and lower time/cost for implementation.

Pollutant Fate and Transport Processes

An additional factor that must be considered is whether the model selected can sufficiently account for the pollutant fate and transport processes needed to address the problem at hand. For example, some models are well suited to handle nitrogen transport but are not well suited to handle phosphorus transport.

Selecting Models that Incorporate the Appropriate Pollutant Transport Mechanisms

Once the appropriate type of model has been chosen for use, the next step is to choose a particular model that can handle the transport and transformation of the pollutants that are of concern. For example, some models do a very good job simulating nitrogen transport but do not have rigorous formulations for phosphorus transport. Some models that can simulate nitrogen and phosphorus transport and transformations cannot handle microbial transport. One of these pollutants may drive the decision-making process. If more than one of these pollutants are considered to be important with regard to the decision-making process, then it is likely that a model with the capability to model multiple pollutants may be selected. This may lead to the selection of a model with simplified formulation but with ability handle multiple pollutants if a robust model that can handle multiple pollutants is not available. However, the results of the simplified model may not be considered reliable and such a model can only be used for preliminary assessment. Thus, the decision to select a model type, in this case, will likely depend on the level of accuracy needed for each pollutant.

Model Parsimony and Transparency

In the development and use of models, parsimony refers to the preference for the least complicated explanation for an observation. Transparency refers to the need for stakeholders and members of the public to comprehend the essential workings of the model and its outputs. Parsimony does not justify simplicity. It instead demands that a model capture all essential processes for the system under consideration in the simplest way possible. It requires that models meet the difficult goal of being accurate representations of the system of interest while being reproducible, transparent, and useful for the regulatory decision at hand (Strimbu et al., 2017).

The need to move beyond simple validation exercises to a more extensive model evaluation leads to the need to explicitly assess the trade-offs that affect parsimony, transparency, and other considerations in the process of developing and applying models. These trade-offs are important to modelers, regulators, and stakeholders. For example, when a physically-based model can accurately define a chemical release to surface water based on a detailed mass balance, should the regulator consider an empirical model that has the same level of accuracy? Here, parsimony might give preference to the simpler empirical model, whereas transparency is best served by the mass balance model that allows the model user to see how the release is transformed into a concentration.

Model Calibration, Validation and Uncertainty Analysis

Introduction

Watershed models are useful tools for simulating the effect of watershed processes and management on soil and water resources. They should pass through calibration and evaluation processes before they are used as a decision-making tool. The important steps include data collection, model setup, parameter identification through prior knowledge and sensitivity, calibration, validation, model evaluation, and model uncertainty. Each of the steps is presented in the subsequent sections. Specific application steps or procedures may differ based on the model type used due to the variations of the specific physical, chemical, and biological systems they each attempt to represent, however, they, in general, have many steps in common.

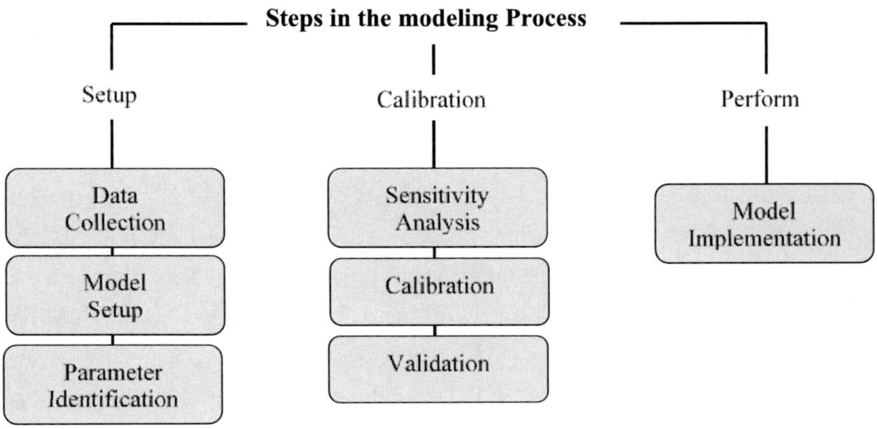

Figure 2. Watershed modeling process.

Figure 2 shows the general steps that are followed in the modeling process. The calibration and validation steps are especially critical since the outcome establishes how well the model represents the watershed for the purpose of the study, although the other steps are also an important part of the process.

Model Setup

The main goal in watershed modeling is to represent the natural system as accurately as possible using data. Thus, watershed models need a diversity of information (land use, soil, geological, topographic, meteorological and hydrological) in order to set up and run. The amount of data required depends on the complexity of the model. Some models can be run using data available in tabular form and may not contain specific spatial or geographical locations. Other models are built from a combination of tabular information and geographically referenced data (i.e., GIS data) or entirely within a GIS framework from datasets that are geographically referenced such as digital elevation models (DEM), land use-land cover data and soils data.

Model Initialization

An initialization (warm-up) period is used to allow the water resources model to run for a sufficient period prior to the simulation period to initialize important model variables or allow important processes to reach a dynamic equilibrium. For example, Muthuwatta et al. (2009) cited an insufficient warm-up period for causing reduced model performance, particularly in the initial years of simulation. Length of the warm-up period may vary for different watershed-scale processes. Warm-up periods in hydrology and water quality studies may range from months to decades, with one to four years being common for watershed-scale modeling. The time needed to reach a realistic initial state is likely related to the temporal and spatial scale of the governing processes. The structural complexity of the model will also impact the length of the warm-up period. Finally, the time needed to approach an equilibrium state will increase the deviation of initial state values relative to their actual values (Daggupati et al. 2015). Model output during the initialization is excluded from the output summary.

Model Calibration

Watershed models simplify field conditions, thus the underlying assumptions and inputs used to influence the model results. Model calibration is the process of estimating model parameters by comparing model predictions (output) for a given set of assumed conditions with observed data for the same conditions.

Watershed models require calibration before they are utilized as a decision-making tool (Geza et al., 2009; Ma et al., 2014; Shojaeizadeh et al. 2018; Shojaeizadeh et al. 2019). Calibration, also referred as inverse modeling, is a process of comparison on an appropriate scale of the simulation results for chosen variables with a set of observations of the variables (Figure 3). During calibration, the differences between the simulated and observed variables are minimized by finding an optimum set of model parameters.

Figure 3 shows a schematic of an inverse analysis procedure. The input parameters are initially estimated by conventional means (e.g., using available laboratory and field test results). The next step is to run the numerical simulation model to generate simulated values. The simulated values are then compared to the field observations, and the regression analysis is performed to minimize an objective function. The objective function quantifies the fit between computed results and observations and it is minimized by updating the set of input parameters needed to perform the numerical simulation. If the model fit is not "optimal", the procedure is repeated until the model is optimized.

Model calibration is used to improve parameter estimates before the model is used for decision-making. It is a complex process because model errors can originate from diverse sources including the limitation in the mathematical structure of the model, limitations of the input data, and lack of knowledge of the watershed characteristics. Calibration requires that we decide which calibration criteria or objective function to use and determine the set of parameters that can meet our calibration criteria. There are various methods of calibration ranging from manual or trial-and-error method to automatic calibration methods using an optimization technique. The manual method is time-consuming and subjective, and does not include a procedure for seeking optimal values, nor provide associated calibration statistics for model analysis. In contrast, "automated" calibration is faster, less subjective, and can provide valuable statistics.

There are quite a few published contributions involving automatic calibration of distributed watershed-scale models. The studies by Senarath et al. (2000), Ajami et al. (2004), and Eckhardt and Arnold (2001) used a Shuffled Complex Evolution (SCE-UA) (Duan et al., 1992) search algorithm. Muleta and Nicklow (2005) used genetic algorithms (GAs) (Holland, 1975) for automatic calibration. Most of these methods used Monte Carlo approaches. In the Monte Carlo method, a model is run multiple times, each time changing the values of the model inputs and parameters by randomly sampling from their associated population distributions. Output values from all simulations are collected, the distribution of values is plotted and analyzed, and the impact of the input variation on the output variation is observed. The output distribution is compared with the distribution of the actual measurements.

There are also other studies that implemented multiple objective optimizations, which improve model calibration but increase the computational complexity. The classical multi-objective optimization involves transforming multiple objectives into a single function. Other studies implemented Pareto optimization to avoid the disadvantages of converting multi-objective functions into a single optimization problem (Boyle et al., 2001; Vrugt et al., 2003; Khu and Madsen, 2005), by generating a set of solutions that provide an equally good fit. Use of a single objective function is appropriate when objectives do not conflict. In this case, a Pareto optimization will converge to a single set of parameter values. Multi-objective calibrations deal mainly with unrelated functions that do not conflict. Gupta et al. (1998) discussed the use of alternative error statistics as objective functions in the multi-objective calibration of a model to simulate one aspect (i.e., daily flow) of a watershed system. They used a general-purpose global multi-objective optimization algorithm (MOCOM-UA) and successfully estimated the Pareto solution space. They noted that the objectives should be unrelated functions that measure different aspects of the relationship between the data and the parameters of the system.

Parameter-estimation involves some kind of criteria (convergence criteria) to select final optimal parameter values: (1) changes in the parameter values, where a unique criterion can be specified for each parameter, and (2) the change in the sum-of-squared-weighted residuals. If the changes are too large and the maximum number of parameter-estimation iterations has not been reached, the next parameter-estimation iteration is executed. If the changes are small enough, parameter estimation converges. If convergence is achieved because the changes in the parameter values or the changes in the objective function (depending which criteria is used) are small enough the parameter values are more likely to be the optimal parameter values – that is, the values that produce the best possible match between the simulated and observed values, as measured using the weighted least-squares objective function.

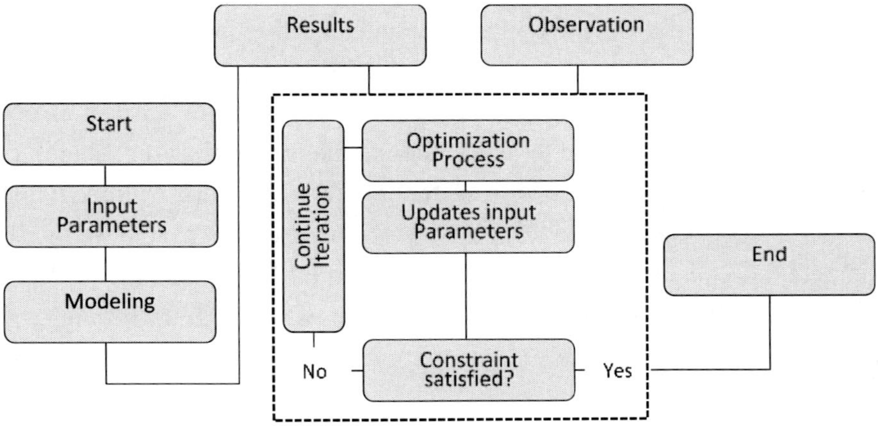

Figure 3. The inverse analysis procedure.

Sensitivity Analysis

Sensitivity analysis is the process of determining the rate of change in model output with respect to changes in model inputs (parameters). Sensitivities are usually defined by computing partial derivatives of the output functions with respect to the parameters. The sensitivity can be calculated for a small change of one parameter while the other input parameters are held constant. Watershed models are calibrated against measured data by adjusting the parameter values according to certain calibration criteria. These models represent the different physical phenomena occurring within the catchment. As a result, proper calibration and validation require a huge amount of data (land use, soil, geological, topographic, meteorological and hydrological) representing the various variables. Therefore, these models involve a large number of model parameters. It is not practical to use all parameters in calibration especially for a distributed model with a large number of parameters. Thus, the number of parameters is reduced either by using prior information to set parameters values within a certain range using information about the watershed, by a parameterization technique or through sensitivity analysis.

It is necessary to identify model parameters that could be reasonably estimated based on field data alone, using prior information, to reduce the number of parameters to be determined via calibration. To this effort, Refsgaard (1997) suggested to assess the parameter values from field data as much as possible and to fix spatial patterns of parameters to simplify the calibration process. Another technique to reduce the dimensionality of the parameter space is to use sensitivity analysis on the model output. Through sensitivity analysis, the parameters that are non-essential in influencing the model response can be fixed to their prior values (Muleta and Nicklow, 2005; Christiaens and Feyen, 2002). Not knowing the sensitivity of parameters can also result in spending too much time on in-sensitive parameters that do not affect the modeling result significantly. Focus on sensitive parameters can lead to a better understanding and to better-estimated values and thus reduced uncertainty. Therefore, sensitivity analysis as an instrument for the assessment of the input parameters with respect to their impact on model output is useful for model calibration and for reduction of uncertainty.

A variety of methods are available for sensitivity analysis. Existing methods can be classified based on the way the parameters are treated (Saltelli et al., 2000). There are both local and global methods of analysis. The local methods estimate the local impact of a parameter value on the model output. Global techniques, on the other hand, analyze the whole parameter space at once. A local sensitivity analysis evaluates sensitivity at one point (e.g., a default value) in the parameter space.

Model Validation

Model validation involves running a model using input parameters measured or determined during the calibration process. Validation procedures are similar to calibration procedures in that both try to match the predicted values to the measured values. However, a dataset of measured watershed response selected for validation preferably should be different than the one used for model calibration, and the model parameters are not adjusted during validation but evaluated to see if they can reproduce good match between predicted and observed values. These limits for adequate calibration are stricter than the rating for general model validation because model parameter values are optimized during calibration but not during model validation or application. Validation provides a test of whether the model was calibrated to a particular dataset or the system it is to represent. If predictions do not produce good results for the validation dataset, calibration and/or model assumptions may be revisited. Generally, the objective of the validation step is to test the performance of a calibrated model parameter set against an independent set of measured data. The measured validation and calibration data sets cover different time periods or conditions. Good validation results support the usefulness of the model to predict future conditions under alternative management scenarios and future climates. Since validation assesses the performance of a calibrated model parameter set against a set of independently measured data, it is typically more difficult to get a good validation performance in comparison to calibration. According to Refsgaard (1997), model validation is the process of demonstrating that a given site-specific model is capable of making "sufficiently accurate" simulations, although "sufficiently accurate" can vary based on project goals. According to the U.S. EPA (2002), the process used to accept, reject, or qualify model results should be established and documented before beginning model evaluation.

Model Evaluation

Model users would like to know how good the calibration results are and how to measure them (goodness-of-fit). Thus, model evaluation is discussed here in the context of model calibration and validation. The techniques discussed are used during model calibration and validation to see if simulation results are close enough to the observed data. There are both statistical and graphical model evaluation techniques that are applied to the calibration and the validation of models. The statistical methods may include standard regression, dimensionless and error-index. Graphical methods can also be used to visually compare measured and observed values to get an overview of model performance. A number of goodness-of-ft measures can be used to evaluate the overall performance of a simulation model, Few of the statistical methods include the slope and y-intercept, the coefficient of determination (R^2), Index of agreement (d), Nash-Sutcliffe efficiency (NSE), and error indices such as Mean Absolute Error (MAE), the Mean Squared Error (MSE), the Root Mean Squared Error (RMSE) and Normalized mean bias (NMB).

The slope and y-intercept of the best-fit regression line can indicate how well-simulated data match measured data. The slope and y-intercept are commonly examined under the assumption that measured and simulated values are linearly related, which implies that all of the error variance is contained in simulated values and that measured data are error-free (Willmott, 1981). The coefficient of determination (R^2) describes the degree of co-linearity between simulated and measured data. R^2 ranges from 0 to 1, with higher values indicating less error variance, and typically values greater than 0.5 are considered acceptable (Santhi et al., 2001, Van Liew et al., 2003). R^2 is oversensitive to high extreme values (outliers) and insensitive to additive and proportional differences between model predictions and measured data (Legates and McCabe, 1999).

The index of agreement (d) developed by Willmott (1981) is another measure of a standardized measure of the degree of model prediction error. It is calculated as:

$$d = 1.0 - \frac{\sum_{i=1}^{N} |Q_O - Q_S|}{\sum_{i=1}^{N} |Q_S - \overline{Q_O}| + |Q_O - \overline{Q_O}|} \quad (1)$$

where; Q_o = measured streamflow, Q_s = simulated flow and $\overline{Q_o}$ is the measured mean streamflow. The index of agreement, d, varies between 0 and 1. A value of 1 indicates a perfect agreement between the measured and predicted values, and 0 indicates no agreement at all (Willmott, 1981).

The Nash-Sutcliffe Efficiency (NSE) (Nash and Sutcliffe, 1970) determines the model efficiency as a fraction of the measured streamflow variance that is reproduced by the model. NSE is given is calculated as:

$$NSE = 1 - \frac{\Sigma(Q_O - Q_S)^2}{\Sigma(Q_O - \overline{Q_O})^2} \quad (2)$$

where: Q_o = measured streamflow, Qs = simulated flow and $\overline{Q_o}$ is the measured mean streamflow. The closer the NSE value to 1.0 the better is the estimation of the streamflow by the model. NSE ≥ 0.75 is considered to be an excellent estimate, and NSE between 0.75 and 0.36, is regarded to be satisfactory (Motovilov et al., 1999).

Error indices including mean absolute error (MAE), mean square error (MSE), root mean square error (RMSE), the ratio of RMSE to standard deviation ratio (RSR), normalized mean bias (NMB) are frequently used model evaluation. RMSE, MAE, MSE, RSR and NMB values of 0 indicate a perfect fit. Singh et al. (2004) state that RMSE and MAE values less than half the standard deviation of the measured data may be considered low and that either is appropriate for model evaluation.

The mean absolute error (MAE) function is given by:

$$MAE = \frac{\sum_{i=1}^{n} |Q_S - Q_O|}{n} \quad (3)$$

where; Q_s = simulated streamflow and Q_o = measured streamflow and n is the number of observations. As the name suggests, the mean absolute error is a weighted average of the absolute errors.

The mean squared error or MSE is one of many ways to quantify the amount by which a model predictions differs from the measured value of the quantity being estimated. MSE is called squared error loss. MSE measures the average of the square of the error where the error is the amount by which the model prediction differs from the measured quantity to be estimated.

Root Mean Square Error (RMSE) given by:

$$RMSE = \sqrt{\frac{\sum_{i}^{n} (Q_S - Q_O)^2}{n}} \quad (4)$$

where: Q_s = simulated streamflow, Q_o = measured streamflow and n = the number of observations. An RMSE value closer to zero indicates a better fit to observed values. RSR is the ratio of RMSE and standard deviation standardizes the RMSE.

Normalized mean bias (NMB) measures the average tendency of the simulated data to be larger or smaller than their observed counterparts (Gupta et al., 1998). Normalized mean bias (NMB) is used as normalization to facilitate a range of values. This statistic averages the difference (model - observed) over the sum of observed values. NMB is a useful model performance indicator because it avoids over-inflating the observed range of values, especially at low concentrations. Normalized mean bias is defined as:

$$NMB = \frac{\sum_1^n (Q_s - Q_o)}{\sum_1^n (Q_o)} * 100 \quad (5)$$

where: Q_o = measured stream flow and Q_s is simulated flow.

Graphical techniques such as time series plots, percent exceedance probability curves, bar graphs, and box plots, can also be used to compare seasonal variations and data distributions. Moriasi et al. (2007) reviewed recommended model evaluation techniques, both statistical and graphical. They reported ranges of values and corresponding performance ratings for the recommended statistic and gave recommendation based on the review results and project-specific considerations for streamflow and transport of sediment and nutrients. Based on this analysis, they recommend that three quantitative statistics, Nash-Sutcliffe efficiency (NSE), percent bias, and the ratio of the root mean square error to the standard deviation of measured data (RSR), in addition to the graphical techniques, be used in model evaluation.

Model Uncertainty

Uncertainty in model outputs is the result of the lack of accuracy in model input parameters and the processes the model attempts to describe. Thus, predictive uncertainties may arise due to simplification of the system, errors in the conceptual model, and inadequate quantity and quality of data. In other words, uncertainty can result from either natural variability, such as by unpredictable septic effluent concentration, evapotranspiration, or by both known and unknown errors in the input data, the model parameters, or the model itself. Imperfect knowledge of the values of parameters associated with these processes constitutes parameter error within a model. Natural variability includes both temporal variability and spatial variability, to which model input values may be subject. For example, the rainfall measured at a weather station within a particular model grid cell may be used as an input value for that cell, but the rainfall may actually vary at different points within that cell and may have a different mean. Errors in the model itself, also termed model structural error, may be the result of the imperfect representation of processes in a model. The model structural error can be reduced through model improvement, whereas natural variability of input data is a property of the natural system and is usually not reducible at the scale being used. Computed values differ from observed values, and the magnitude and frequency of these differences characterize the uncertainty of the model estimates.

Parameter values can never be completely accurate. There are also stochastic variables over which we don't have any control. Model simulations are usually performed with a single-point estimate of input variables (e.g., a single denitrification rate for a catchment or soil layer) to give a single deterministic output value (e.g., nitrate concentration). Often, the variability in the data used in the models or the inaccurate representation of the real world by the models is overlooked. In reality, these point estimates belong to distribution that reflects the uncertainties in the data and models used.

Predictions are usually the ultimate purpose of a modeling study. Even models of relatively low predictive accuracy can be useful to decision-makers if the predictive accuracy is appropriately quantified. It is, therefore, essential that modeling efforts be accompanied by an analysis of predictive uncertainty. Uncertainty analysis is used to examine how the lack of knowledge in model parameters, variables, and processes represented by equations propagates through the model structure as model output or forecasting error. Uncertainty of model predictions can be characterized using Monte Carlo simulation, inferential statistics, or regression. Monte

Carlo simulation involves multiple runs of the simulation model each time changing the values of the model inputs and parameters by randomly sampling from their associated population distributions. After finishing the simulations, simulations that are deemed to sufficiently match the field observations are considered. It is assumed that gathering a large number of such model simulations, the distribution of predictions will reflect the prediction uncertainty.

The parameter distributions can be sampled randomly or preferentially and weighted (e.g., Latin hypercube sampling, LHS). Both random sampling and Latin hypercube sampling can be used in the generation of the mapping between analysis inputs and analysis results. In random sampling, a random value is sampled from each distribution specified for each uncertain model parameter. The random sampling approach assumes that the model input variables and model parameters are random. Distributions of these input variables and parameters are defined using expert judgment and by calibration exercises. Once defined, the actual values of these random variables and parameters are drawn from these distributions for each simulation time step. Many simulations are performed, each using values drawn from these probability distributions, to generate a distribution of output values. After simulating a considerable number of time steps, one has a set of equally likely outputs that define a probability distribution for each selected performance measure. Unlike random sampling, Latin Hypercube Sampling (LHS) is a more structured sampling technique. In order to generate samples from input distributions, the range of probable values for each uncertain input parameter is divided into M segments of equal probability. If there are N parameters, there will be M^N possible random samples (combinations). The model is run for each random sample the outputs are analyzed statistically, for example for complementary cumulative distribution function (CCDF) that defines a probability of exceedance. Typically on the order of tens of thousands of model simulations are required to capture the predictive uncertainty.

One example that uses the Monte-Carlo approach is the Generalized Likelihood Uncertainty Estimation (GLUE) method (Beven and Binley, 1992; Beven and Freer, 2001) which is an extension to Spear and

Hornberger's (1980) Generalized Sensitivity Analysis (GSA). In both GLUE and GSA methods ensembles of model parameters are sampled from distributions, typically with independent uniform or normal distributions for each parameter. The model is then run with many such parameter sets, producing multiple sets of model output. These are used together to generate credible regions for model predictions. The generated model parameters can be grouped parameters that produced results consistent with the observed values and parameters that produced results that contradict the observations. These methods are based upon Monte-Carlo simulation. Parameter sets are sampled from a probability distribution, with most reported applications sampling from uniform distributions (Beven, 2001). Each parameter set is used to produce a model output; the acceptability of each model run is then assessed using a goodness-of-fit criterion. Muleta and Nicklow (2005) also used GLUE (Beven and Binley, 1992; Beven and Freer, 2001), which is based on Monte Carlo simulations. A major drawback of the GLUE methodology is the subjectivity of the likelihood level assignment that is used to group the parameter sets into the acceptable and non-acceptable (Muleta and Nicklow, 2005) categories. Lin and Radcliffe (2006) used the Parameter Estimator (PEST) software (Doherty, 2004) to estimate predictive uncertainty. The automatic model calibration and predictive uncertainty analysis based on a Monte Carlo approach were undertaken using a suite of utility codes written to support the use of PEST in the surface water modeling context. Monte Carlo based approaches for evaluating predictive uncertainty require many simulations to obtain a satisfactory description of the output distribution and are typically more time consuming than the use of inferential statistics.

Inferential statistics is another approach to estimate uncertainty. If the model is fairly linear with respect to the parameters, predictive uncertainty can be estimated with only on the order of tens of model runs by using inferential statistics to calculate linear confidence intervals on predictions. Linear confidence intervals are symmetrical with respect to the predicted value. If the model is non-linear, then nonlinear regression can be used to calculate the nonlinear confidence intervals on the predictions, increasing the required number of model runs by twice the number of predictions for which uncertainty is evaluated.

CONCLUSION

The chapter provides an overview of key factors that should be considered in the development and implementation of watershed models. Models, when applied properly, provide useful information for decision-making. Watershed models are frequently applied to evaluate the relative impact of natural and anthropogenic changes on surface water and groundwater quality at local and watershed-scales. Watershed models vary with respect to their capabilities and degree of complexity. Some have multiple component modules with ability to run hydrologic, water quality, and cost benefit analysis while others can handle only some of these processes. Some are developed for specific locations while others are applicable nationwide. Thus, there is a need to explicitly assess the trade-offs that affect parsimony, transparency, and other considerations in the process of selection, development and application of models. Watershed models require calibration before they are utilized as a decision-making tool. These trade-offs are important to modelers, regulators, and stakeholders. Model calibration is used to improve parameter estimates before the model is used for decision-making. Model parameter sensitivity analysis is an important process in the watershed modeling for identifying the most relevant model parameters that influence model out outputs. It is the process of determining the rate of change in model output with respect to changes in model inputs (parameters). Validation procedures are similar to calibration procedures in that both try to match the predicted values to the measured values. A validation approach is used to demonstrate if a given hydrologic model is capable of making sufficiently accurate predictions outside of the calibration period or calibration dataset using the same parameter values estimated via calibration. The model is validated if its accuracy in the validation period is proven to lie within acceptable limits. Discrepancies between observed and simulated values in the validation period may arise from errors in the input data or weaknesses in model structure. A number of goodness-of-fit measures are available to evaluate the overall performance of a simulation model during calibration and validation process. The most common statistical measures include the coefficient of determination (R^2), Nash-Sutcliffe efficiency (NSE), the Mean Squared Error (MSE), the Root Mean Squared Error (RMSE) and Normalized mean bias (NMB). Model uncertainty can result from either natural variability, such as by unpredictable pollutant loading, evapotranspiration, or by both known and unknown errors in the input data, the model parameters, or the model itself. Predictions produced by a calibrated model are more useful when they are reported with an evaluation of uncertainty.

REFERENCES

Ajami, N. K., Gupta, H., Wagner, T. & Sorooshian, S. (2004). "Calibration of a semi distributed hydrologic model for streamflow estimation along a river system", *Journal of hydrology*, *298*, 112-135.

Beven, K. J. & Binley, A. M. (1992). The future of distributed models: model calibration and uncertainty prediction. *Hydrological Processing*, *6*, 279–298.

Beven, K. J. & Freer, J. (2001). Equifinality, data assimilation, and uncertainty estimation in mechanistic modeling of complex environmental systems using the GLUE methodology. *Journal of Hydrology*, *249*, 11–29.

Borah, D. K., Yagow, G., Saleh, A., Barnes, P. L., Rosenthal, W., Krug, E. C. & Hauck, L. M. (2006). *Sediment and nutrient modeling for TMDL development and implementation: Transactions of the ASABE*, v. *49*, no. 4, p. 967-986.

Boyle, D. P., Gupta, H. V., Sorooshian, S., Koren, V., Zhang, Z. & Smith, M. (2001). Toward improved stream flow forecast: value of semi distributed modeling. *Water Resources Research*, *37*(11), 2749–2759.

Christiaens, K. & Feyen, J. (2002). *Use of sensitivity and uncertainty measures in distributed hydrological modeling with an application to the MIKE SHE model, Water Resour. Res., 38*, 1169, doi:10.1029/ 2001WR000 478.

Clement, T. P. (1997). RT3D - A Modular Computer Code for Simulating Reactive Multi-Species Transport in 3-Dimensional Groundwater Aquifers. Pacific Northwest National Laboratory, Richland, WA, USA. PNNL-11720. Found online at: http://bioprocess.pnl.gov/ rt3d.htm.

Clement, T. P., Sun, Y., Hooker, B. S. & Petersen, J. N. (1998). Modeling multi-species reactive transport in ground water. *Groundwater Monitoring and Remediation*, *18* (2), 79–92.

Daggupati, P., Pai, N., Ale, S., Douglas-Mankin, K. R., Zeckoski, R. W., Jeong, J., Parajuli, P. B., Saraswat, D. & Youssef, M. A. (2015). A recommended calibration and validation strategy for hydrologic and water quality models. *Transactions of the ASABE*, *58*(6), pp. 1705-1719.

Danish Hydraulic Institute (DHI). (1998). MIKE SHE Water Movement - Chapter and Technical Reference Manual, Edition 1.1.

Doherty, J. (2004). PEST—Model-independent parameter estimation user's manual. 5th ed. Watermark Numerical Computing, Brisbane, Australia.

Donigan, A. S., Imhoff, J. C., Bicknell, B. R. & Kittle, J. L. (1984). Application guide for hydrological simulation program: FORTRAN (HSPF), U.S. EPA-600/3-84-065, p. 177.

Duan, Q., Sorooshian, S. & Gupta, V. K. (1992). Effective and efficient global optimization for conceptual rainfall–runoff models. *Water Resources and Research*, *28* (4), 1015–1031.

Eckhardt, K. & Arnold, J. G. (2001). Automatic calibration of distributed catchment model. *Journal of Hydrology*, *251*, 103–109.

EPA. (2002). Onsite wastewater treatment systems manual. In: O. Cincinnati, Office of Water, Office of Research Development (Editor).

Gaines, A. G. (1986). Lagoon Pond Study: An Assessment of Environmental Issues and Observations on the Estuarine System. Final Report for the Town of Oak Bluffs. Woods Hole Oceanographic Institution, Woods Hole, Massachusetts, USA.

Geza, M. & McCray, J. E. (2008). Effects of soil data resolution on SWAT model stream flow and water quality predictions. *Journal of environmental management*, *88*(3), pp. 393-406.

Geza, M., Poeter, E. P. & McCray, J. E. (2009). Quantifying predictive uncertainty for a mountain-watershed model. *Journal of hydrology*, *376*(1-2), pp. 170-181.

Ghanem, A., Steffler, P. M., Hicks, F. E. & Katopodis, C. (1995). "Two dimensional finite element model for aquatic habitats", Water Resources Engineering Report 95-S1, Dept. of Civil Engineering, University of Alberta, 189 pp.

Green, W. H. & Ampt, G. (1911). Study on soil physics-1. The flow of air and water through soils. *J. Agric. Sci.*, *4*, 1–24.

Gupta, H. V., Sorooshian, S. & Yapo, P. O. (1998). Toward Improved Calibration of Hydrologic Models: Multiple and Noncommensurable Measures of Information. *Water Resources Research*, *34*(4), 751-763.

Herr, J. L. Weintraub. & Chen, C. W. (2001). User's Guide to WARMF. Documentation of Graphical User Interface. Prepared for Electric Power Research Institute, Palo Alto, California.

Hill, M. C. (1992). A Computer Program (MODFLOWP) for Estimating Parameters of a Transient, Three-dimensional, Groundwater Flow Model Using Nonlinear Regression. Open File Report, 91-484, USGS.

Holland, J. H. (1975). Adaptation in Natural and Artificial Systems. University of Michigan Press, Ann Arbor, Michigan., 183 pp.

Holtan, H. N., England, C. B & Shanholtz, V. O. (1967). Concepts in Hydrologic Soil Grouping. *Transactions of the ASAE*. Paper No. 65-739, 407- 410.

Horton, R. E. (1939). Analysis of runoff-plat experiments with varying infiltration-capacity. Eos, *Transactions American Geophysical Union*, *20*(4), pp. 693-711.

Hydrologic Engineering Center. (1995). HEC-RAS River Analysis System - User's Manual, US Army Corps of Engineers, Davis, CA.

Khu, S. T. & Madsen, H. (2005). Multiobjective Calibration with Pareto Preference Ordering: An Application to Rainfall-Runoff Model Calibration. *Water Resources Research*, *41*(3), W03004.

Legates, D. R. & McCabe, G. J. (1999). Evaluating the use of "goodness-of-fit" measures in hydrologic and hydroclimatic model validation. *Water Resources Res.*, *35*(1), 233-241.

Lin, Z. & Radcliffe, D. E. (2006). Automatic Calibration and Predictive Uncertainty Analysis of a Semi distributed Watershed Model. *Vadose Zone Journal*, *5*, 248-260.

Ma, G., Geza, M., & Xu, P. (2014). Review of Flowback and Produced Water Management, Treatment, and Beneficial Use for Major Shale Gas Development Basins. In Shale Energy Engineering 2014: Technical Challenges, Environmental Issues, and Public Policy (pp. 53-62).

Maxwell, R. M. & Miller, N. L. (2005). Development of a coupled land surface and groundwater model. *Journal of Hydrometeorology*, *6*(3), pp. 233-247.

Moriasi, D. N., Arnold, J. G., Van Liew, M. W., Bingner, R. L., Harmel, R. D. & Veith, T. L. (2007). Model evaluation guidelines for systematic quantification of accuracy in watershed simulations. *Transactions of the ASABE.*, *50*(3), 885-900.

Motovilov, Y. G., Gottschalk, L., England, K. & Rodhe, A. (1999). Validation of a distributed hydrological model against spatial observations. *Agric. and Forest Meteorology*, 98-99, 257-277.

Muleta, M. K. & Nicklow, J. W. (2005). Sensitivity and uncertainty anaylsis coupled with automatic calibration for a distributed watershed model. *J. of Hydrology*, *306*, 127-145.

Muthuwatta, L. P., Booij, M. J., Reintjes, T. H., Bos, M. G., Geiske, A. S. & Ahmad, M. (2009). Calibration of a semi-distributed hydrological model using discharge and remote sensing data. In *New Approaches to Hydrological Prediction in Data-Sparse Regions*, (pp. 52-58). IAHS Publication 333. Wallingford, U.K.: 1718 TRANSACTIONS OF THE ASABE International Association of Hydrological Sciences.

Nash, J. E. & Sutcliffe, J. V. (1970). River flow forecasting through conceptual models, part 1 - A discussion of principles, *J. of Hydrology*, *10*(3), 282-290.

Refsgaard, J. C. (1997). Parameterization, calibration and validation of distributed hydrologic models. *Journal of Hydrology*, *198*, 69–97.

Richards, L. A. (1931). Capillary conduction of liquids through porous mediums. *Physics*, *1*(5), pp. 318-333.

Saltelli, A., Chan, K. & Scott, E. M. (Eds.). (2000). Sensitivity Analysis. Wiley, New York.

Santhi, C., Arnold, J. G., Williams, J. R., Dugas, W. A., Srinivasan, R. & Hauck, L. M. (2001). Validation of the SWAT model on a large river basin with point and nonpoint sources. *J. American Water Resources Assoc.*, *37*(5), 1169-1188.

SCS. (1972). Section 4: Hydrology, National Engineering Handbook: Washington, DC, USDA Soil Conservation Service.

Senarath, S. U. S., Ogden, F. L., Downer, C. W. & Sharif, H. O. (2000). On the calibration and verification of two-dimensional, distributed, hortonian, continuous watershed models. *Water Resources and Research*, *36* (6), 1495–1510.

Shojaeizadeh, A., Geza, M., Bell, C., Gallo, E., Spahr, K., Hogue, T., & McCray, J. (2018). A Site Scale Integrated Decision Support Tool for Urban Stormwater Management. In World Environmental and Water Resources Congress 2018: Water, Wastewater, and Stormwater; Urban Watershed Management; Municipal Water Infrastructure; and Desalination and Water Reuse (pp. 112-120). Reston, VA: American Society of Civil Engineers.

Shojaeizadeh, A., Geza, M., McCray, J., & Hogue, T. S. (2019). Site-Scale Integrated Decision Support Tool (i-DSTss) for Stormwater Management. Water, 11(10), 2022.

Simunek, J., Sejna, M. & Van Genuchten, M. T. (1999). The HYDRUS-2D software package for simulating two-dimensional movement of water, heat, and multiple solutes in variably saturated media. Version 2.0, International Ground Water Modeling Center: Golden, CO, p. 251.

Singh, J., Knapp, H. V. & Demissie, M. (2004). Hydrologic modeling of the Iroquois River watershed using HSPF and SWAT. ISWS CR 2004-08. Champaign, Ill.: Illinois State Water Survey. Available at: www.sws.uiuc.edu/pubdoc/CR/ ISWSCR2004-08.pdf.

Spear, R. C. & Hornberger, G. M. (1980). Eutrophication in Peel Inlet, II, Identification of critical uncertainties via generalized sensitivity analysis. *Water Research*, *14*, 43–49.

Strimbu, B. M., Amarioarei, A. & Paun, M. (2017). A parsimonious approach for modeling uncertainty within complex nonlinear relationships. *Ecosphere*, *8*(9).

USDA Soil Conservation Service. (1972). National engineering handbook, Section 4: Hydrology, Chapters 4–10, Washington, DC., 15-7–15-11.

Valiela, I., Bowen, J. L. & Kroeger, K. D. (2002). *Assessment of models for estimation of land derived nitrogen loads to shallow estuaries: Applied Geochemistry*, v. *17*, no. 7, p. 935-953.

Valiela, I., Collins, G., Kremer, J., Lajtha, K., Geist, M., Seely, B., Brawley, J. & Sham, C. H. (1997). *Nitrogen loading from coastal watersheds to receiving estuaries: New method and application: Ecological Applications*, v. *7*, no. 2, p. 358-380.

Van Liew, M. W., Arnold, J. G. & Garbrecht, J. D. (2003). Hydrologic simulation on agricultural watersheds: Choosing between two models. *Trans. ASAE*, *46*(6), 1539-1551.

Vrugt, J. A., Gupta, H. V., Bastidas, L. A., Bouten, W. & Sorooshian, S. (2003). Effective and Efficient Algorithm for Multiobjective Optimization of Hydrologic Models. *Water Resources Research*, *39*(8), 1214.

Willmott, C. J. (1981). On the validation of models. *Physical Geography*, *2*, 184-194.

Zheng, C. (1993). A modular three-dimensional transport model for simulation of advection, dispersion and chemical reaction of contaminants in groundwater systems. United States Environmental Protection Agency Report.

Zheng, C. & Wang, P. P. (1999). MT3DMS, A modular three-dimensional multi-species transport model for simulation of advection, dispersion and chemical reactions of contaminants in groundwater systems; documentation and user's guide, U.S. Army Engineer Research and Development Center Contract Report SERDP-99-1, Vicksburg, MS, 202 p.

In: Sustainable Water
Editors: V. Gnaneswar Gude et al. ISBN: 978-1-53617-335-2

Chapter 3

COMPUTER TOOLS FOR URBAN HYDROLOGY AND WATER QUALITY MANAGEMENT

Jejal Reddy Bathi **and** ***Shuvashish Roy***

Civil and Chemical Engineering,
University of Tennessee at Chattanooga,
Chattanooga, Tennessee, US

ABSTRACT

Hydrology of urban watersheds is different from natural watersheds due to their higher impervious area, lesser vegetation, and decreased evapotranspiration. Besides, anthropogenic pollutants like polycyclic aromatic hydrocarbons from parking lots, roads, fossil fuel byproducts and heavy metals from industry areas contribute to high levels of pollution in urban runoff. Considering the ever-growing population in the urban areas, sustainable water resources management approaches are critical not only to minimize the impact on the environment and public health but also to confirm the adequacy of water supply. To assist with sustainable water resources planning, numerous numerical computer programs (computer models) have been developed that perform a series of interactive,

* Corresponding Author's Email: jejalreddy-bathi@utc.edu.

unvarying steps to define existing conditions, analyze problems, and to help investigate possible solutions. However, with an increasing number of models that often have similar or the same capabilities for watershed water resources analysis, users are often overwhelmed when selecting the best model that suits their requirements. In order to introduce readers to watershed simulation computer programs evolution, in this chapter, a historical perspective of urban watershed models alongside new developments is provided. A detailed discussion of selected popular watershed models, classified into two major categories (simple models and complex models), is provided to inform the readers of the features and limitations of those models. Critical summary of a variety of watershed simulation models is presented in a tabular format which can be used as a screening tool for side by side comparison of prominent models and hence help choose a right model for the purpose.

Keywords: watershed modeling, urban stormwater models, hydrology, runoff quality, hydraulics

INTRODUCTION

Urbanization changes the land use patterns and adversely influences the natural hydrological processes [1, 2]. Lack of natural terrain due to the expansion of built area seriously limits the stormwater infiltration capacity, which exacerbates surface runoff and increases flood peaks and volumes [3-5]. Besides, urbanization has been reported to degrade stormwater quality with the presence of higher sediment and nutrient loads [6, 7]. Urban flooding and stormwater pollution are manmade synthetic problems, which is significantly different from rural/natural flooding, thus requires an engineered approach to mitigate economic, health and social disruptions.

Urban storm systems in the earlier generations primarily focused on water quantity aspects and emphasized the goal of draining stormwater and wastewater from the urban areas in a rapid manner. The modern urban stormwater systems have evolved as a robust and integrated system that address both water quantity and quality issues. Design of such modern systems often relies upon computer technology equipped with engineering and water resources sciences that facilitate faster and accurate analysis of hydrology and water quality aspects of urban infrastructure. These computer technologies, i.e., watershed models, are often required to address anticipated needs and unexpected problems associated with urban flooding and degrading urban surface water quality. With the rapid advancement in computer technology and the recognized need for computer models in urban water resource management, dozens of commercial and open-source urban watershed models are developed and are now available for users. Such models are useful to study the complex hydrological activities and their impacts on the urban areas and develop subsequent urban watershed controls for mitigating negative impacts of urbanization.

In the 1850s, especially in Europe, surface runoff was estimated based on empirical results. At that time the engineers used Roe's Table to size sewer pipes for draining a specified size catchment [8]. Mulvaney (1851) and Kuichling (1889) developed a rational method for hydraulic and hydrologic design. These methods became popular in later days for hydraulic and hydrology planning and design and even now they are still in vogue [9]. In the late 1960s, computer-aided modeling of stormwater runoff was introduced for the first time by the United States Environmental Protection Agency (EPA) followed by many other mathematical models developed thereafter [8]. In the 1970s, the urban watershed modelers incorporated aspects of both water quality and water supply [10]. In general, most of the early models are based on conventional methods for runoff generation and routing. Some of the early models have been updated with groundwater and baseflow components and several others with infiltration components [11]. Increased use of personal computer and different pre and post processors like Geographical Information System (GIS) in the late 1990s has dramatically improved the computer models and hence stormwater management practices for urban areas.

Over the period, many modern models emerged from different academic and commercial organizations. Most of them can simulate both single event and multiple event runoff quantities and quality with varying spatial and temporal resolutions. These models primarily differ from each other on the basis of their approach, functionality, accessibility and hydrology and water quality components. Predicted accuracy of a model depends on the extent of details of hydrology and water quality algorithms incorporated in the individual models. Even with significant advancement in the modeling approach and model capabilities, there is no model that is accurate; however, some models have better accuracy than the others. For example, a review by Obropta and Kardos [12] concluded that the models with hybrid approaches, integrating deterministic and stochastic models, are capable of simulating both water quality and quantity and show strong potential for reducing prediction error & uncertainty in water quality analysis compared to the deterministic and stochastic approaches alone.

The modern-day computer models are now geared towards the sustainability of urban water systems to combat increasing stress on the urban waters in terms of quantity and quality. While urban watershed models can be used for many purposes, typical applications of urban watershed models include:

- Total Maximum Daily Load (TMDL)
- Flood plan mapping
- Flood Forecasting
- Peak flow analysis
- Sediment and contaminant analysis
- Thunderstorm flood analysis
- Surface ponding and infiltration analysis
- Impact of land cover and land use change analysis
- Modeling of wetland
- Planning, design, and operation of hydraulic structures
- Storm drain modeling
- Groundwater-surface water interactions
- Pollutants fate and transport analysis
- Most recent models are increasingly used for water harvesting and re-use analysis

Considering the rapid progress in science and computational technology, one can anticipate further modification and recalibration of the urban stormwater management models in the near future. Potential areas of improvements in urban stormwater modeling include: improvement of rainfall-runoff and water quality simulation parameters, reduction of uncertainties, broadening the range of Low Impact Development (LID) and Green Infrastructure (GI) features, better approximation in simulating non stationary and nonlinear time series data, real-time model calibration, creating linkage with different models i.e., habitat to toxicity models, algorithm efficiency improvement, and application of artificial intelligence [13].

In summary, knowledge of watershed management tools is crucial for the planners and decision-makers for adopting cost-effective and sustainable decision making for improved water quality and quantity management. The goal of this book chapter is to provide a review of prominent urban watershed models that can help readers to gain knowledge on available models and their capabilities and hence make improved decisions in selecting an appropriate model for their purpose.

Classification of Urban Watershed Models

Chow, Maidment [14] classified hydrological models into physical and mathematical models. Physical models represent experimental models while the mathematical models involve empirical or conceptual equations of chronological set of relation, numerical and logical steps [15]. Physical models can either be a downsized model of a genuine framework or an analog model which is the result of a simulated process representing a natural or target process.

Mathematical models can be further classified as deterministic and stochastic models. A deterministic model produces the same output results for identical input parameters while the stochastic models produce different results for the same input parameters due to the use of random variables with the probabilistic distribution. Natural process (i.e., rainfall, snowfall, humidity) are often impossible to predict with hundred percent certainty and thus follows a probabilistic distribution. Stochastic models use the probabilistic variables to resemble the uncertainty of the natural processes. While the deterministic model makes a forecast based on the input parameters, a stochastic model creates a prediction based on the randomness of variables in terms of time and space [16]. Because of this randomness, the stochastic models often yield inaccurate and inconsistent results [17].

Zoppou reviewed several urban watershed models and presented them according to their functionality, accessibility, water quantity and quality components included in the models [18]. Urban watershed models can also be classified as traditional and integrated models. While the traditional models are primarily equipped for hydraulic analysis, more recent trends of integrated models are capable of simulating both stormwater quality and quality.

Integrated models include surface, vadose zone, groundwater, and even atmospheric component. Pitt and Burton [19] classified stormwater catchment models based on their complexity. Watershed models can also be classified in terms of temporal and spatial resolutions. Certain models such as Mike Urban and UrbanCycle are capable of running at the sub-hourly temporal resolution, whereas others run only at a daily time step, which is a serious limitation to represent the fast hydrological response of urban watersheds [10, 20]. As a result, models with daily time steps are losing popularity when compared to the sub-hourly models [10].

Mathematical models can also be classified into physically based or conceptual and empirical models. A model based on physical laws like Darcy's law and Newton's law of motion is regarded as a conceptual model [18]. Nix [21] divided the models as distributed and lumped models based on the spatial variability of the input parameters. A lumped model takes average values of the spatial distribution of the input, whereas distributed models consider spatial variability across the domain. In the distributed model the watershed area is divided into smaller units. A lumped model is applied in each unit of a distributed system. Most urban runoff models are deterministic-distributed models [18].

Urban watershed models can also be classified into hydrological, hydraulic and water quality models. Hydrologic models are focused on hydrologic cycle and are thus capable of runoff estimation and channel routing from drainage area based on rainfall intensity, topography, etc. Hydraulic models are more related to the flow of water through the hydraulic structures (i.e., river, pipe, etc.). Models capable of simulating pollutant loading to surface waters or pollutant removal in a water quality treatment unit are getting popularity for urban watershed modeling. Thus, many stormwater models are adding features for water quality controls which is essential for sustainable urban water resource management.

The watershed models can again be classified based on temporal resolution (time series or time steps), event-based and continuous models. Event-based models are focused on short-term events such as simulating a few or individual storms. These types of models are used for flood forecasting or to assess high flow rate events. Continuous models simulate the hydrological process over a long period of time, involving daily, monthly and seasonal predictions.

Continuous models provide a more realistic representation of the scenario and form the basis of a planning model for urban stormwater management. Continuous models also consider the slow runoff processes such as groundwater flow. Event-based models are regarded as appropriate for infrastructure design as they predict the extreme event (worst-case scenario at hourly or sub-hourly time step). However, wrong or overestimated initial values in the event-based model may provide overfitting or extremely high flood peaks which will lead to over-design of the infrastructure.

Overall, day by day an increasing number of hydrological models are being developed that are based on similar physical and mathematical formulation and function [22]. Moreover, often those models primarily serve a similar purpose. While the use of some of the models is relatively easy in terms of data needs and model set-up, other models are highly data-intensive and complex in setting-up the model.

Thus, the classification of these models by their unique characteristics and application features enables users to identify their usability based on the needs [23].

The models can be classified as simple tools and complex tools. The simple tools can be represented by a simple spreadsheet calculator while the complex tools often involve multiparameter simulation components. In general, simple tools need fewer parameter inputs whereas complex tools require a large number of parameters.

Literature summary of the urban watershed model's classification is illustrated in Figure 1.

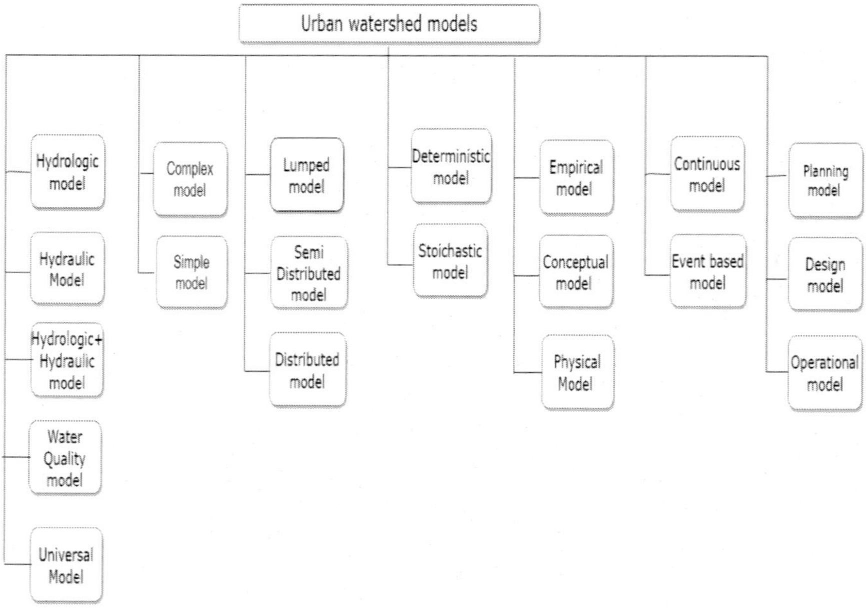

Figure 1. Classification of urban watershed models.

MODEL DESCRIPTIONS

This section provides a description of selected urban watershed models that are grouped into two categories, namely, simple models, and complex models. Models are grouped into each category based on their complexity of use and modeling details. Figure 2 illustrates different urban watershed models, by category, that are discussed in the section. Complex models are discussed in more detail than the simple models as complex models are most commonly applied for planning and designing urban water resources management infrastructure.

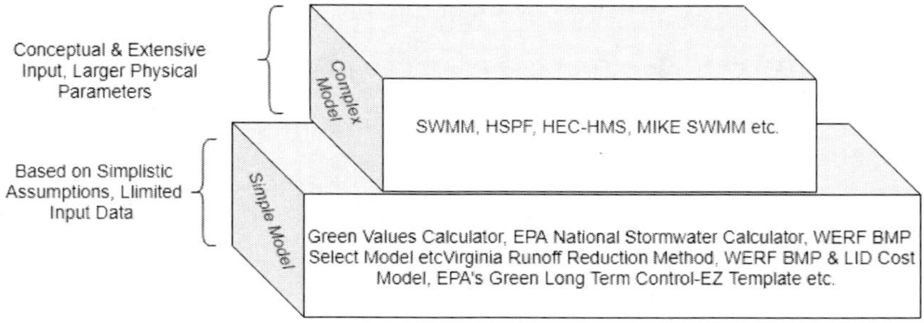

Figure 2. Categories of urban stormwater management models.

Simple Models/Tools

Simple models are screening and primarily planning level tools. These models are useful when the users have fewer resources (i.e., expertise & watershed input data) available. Examples of these models include the U.S. EPA National Stormwater Calculator (SWC) and the Green Values Stormwater Calculator.

National Stormwater Calculator is developed by US EPA with the primary aim to support local, state, and national stormwater management objectives and regulatory efforts to reduce runoff through infiltration and retention using green infrastructure practices as LID controls. EPA SWC applies the Storm Water Management Model (SWMM) at a very basic level, but SWMM has the capacity to simulate extremely complex scenarios in detail when operated independently. It accesses several national online databases such as National Oceanic and Atmospheric Administration (NOAA) and the Soil Survey Geographic Database (SSURGO) to automatically retrieve information (i.e., soil, topography, rainfall, evaporation, etc.) needed as inputs. EPA SWC operates by rerouting the user input information and processing it through EPA's Stormwater Management Model (SWMM). The EPA SWC does not make any actual independent calculations outside of SWMM but acts as an alternative user-friendly interface for users unfamiliar with hydraulic modeling program. The primary focus of the SWC is to help assess how well a desired stormwater retention target can be met with and without the use of green infrastructure.

SWC can model runoff, sediment and nutrient transport based on local soil type, land cover, weather/rain data, topography, etc. It also has options for analyzing climate change scenarios. The National Stormwater Calculator allows users to apply any mix of LID controls by simply selecting what percentage of the impervious area is routed through each type of control. The LID controls included in the model are a rooftop disconnection, rainwater harvesting, rain gardens, green roofs, street planters, infiltration basins, and porous pavements controls. The model produces estimates of runoff reduction statistics, percentage of rainfall controlled, maintenance and installation cost of selected LIDs.

Green Values Stormwater Calculator developed by the center for neighborhood technology is a goal-oriented model that provides numerous city and state standards available for selection to streamline runoff reduction goal input. This model can be used to compare a conventional storm sewer system to a green infrastructure-based system. It can estimate annual runoff from a site based on local soil conditions, land cover, historic rainfall records. It also provides cost estimates of planned stormwater controls.

Virginia Runoff Reduction Method, Water Environment Research Foundation (WERF) Best Management Practice (BMP) and Low Impact Development (LID) Whole Life Cost Model, EPA's Green Long-Term Control - EZ Template are other examples of simple models that are spreadsheet tools. Again, these tools enable planning level understanding of stormwater controls such as green roof, downspout disconnection, permeable pavement, cisterns, grass channel, swales, rain barrels, rain gardens, dry swale, bioretention, infiltration, extended detention pond, sheet flow to filter, wet swale, constructed wetland, and wet pond. For the Virginia Runoff Reduction Method, the input requirements include annual precipitation, land cover distribution, soil type distribution, type of control, and the model can provide output results for runoff volume reduction (ft^3 /design storm), phosphorus load reduction (lb/yr) and nitrogen load reduction (lb/yr). The WERF BMP and LID Whole Life Cost Models require drainage area, control characteristics, capital costs, maintenance costs as inputs and the model provide whole life costs of the controls as output. The U.S. EPA's Green Long-Term Control - EZ Template is useful for integrated green infrastructure planning for combined sewer overflows control by planning to target reduced stormwater runoff volume and hence combined sewer overflow (CSO) volume. It also provides cost estimates of planned stormwater controls.

Complex Models

Complex models are primarily conceptual because they are based on physical laws. As they involve a larger number of physical parameters, input data requirements are often very extensive. The most commonly used complex models that include: EPA Storm Water Management Model (SWMM), Hydrological Simulation Program-Fortran (HSPF), Hydrologic Modeling System by the Hydrologic Engineering Center (HEC-HMS), Windows-based Technical Release 55 (Win TR-55), MIKE SWMM by Danish Hydraulic Institute (DHI), Urban Water Volume and Quality (UVQ), Model for Urban Stormwater Improvement Conceptualisation (MUSIC), STORM, Source Loading and Management Model for Windows (WinSLAM), System for Urban Stormwater Treatment and Analysis Integration (SUSTAIN), Gridded Surface Subsurface Hydrologic Analysis (GSSHA), and Watershed Management System (WMS) are discussed here.

Storm Water Management Model (SWMM)

SWMM is a widely used rainfall-runoff simulation model primarily developed for urban runoff simulation by the U.S. EPA as SWMM-I in 1971 with the affiliation of water resources engineers of the University of Florida. Since then, it has gone through several major updates. The latest Windows version of the model is SWMM 5.1. It is a dynamic hydraulic and hydrologic model with capabilities to simulate runoff and pollutant loads in a collection of sub-areas for a short term (single event) or long term (continuous event) basis [24]. Its newly added graphical user interface comes with study area inputs editing option, hydrologic, hydraulic and water quality simulation modules, and multiple output options including series of the color-coded drainage area and conveyance maps, time series graphs and tables, profile plots, and statistical frequency analysis.

SWMM model estimates surface runoff based on rainfall intensities, antecedent moisture conditions, land use, and topography. It applies conservation of mass, energy or momentum to simulate runoff quantity with the help of simple nonlinear reservoir storage. Infiltration is routed through upper and lower subzones and may return as runoff. SWMM offers three choices for modeling infiltration, Horton's Equation, Green Ampt Method and Curve Number Method. Horton's method assumes infiltration decreases exponentially over the course of a long-term rainfall event. Horton's method requires the maximum, minimum infiltration rates, decay rate coefficient for infiltration and the time over which the fully saturated soil becomes completely dry. Green Ampt method divides the soil column into two separate zones by a wetting front; a soil with some initial moisture content on the bottom and a saturated soil zone on the top of it. This method requires initial moisture deficit of the soil, the soil's hydraulic conductivity, and the suction head at the wetting front as inputs. On the other hand, the Curve Number Method adopts the infiltration capacity of the soil from the soils tabulated curve number (NRCS). The input parameters for this method are the curve number, the soil's hydraulic conductivity and the time over which the fully saturated soil becomes completely dry. In this way, it can be used to reduce the effect of soil imperviousness by altering these infiltration parameters and thus can improve soil protection and flooding impact. Flow routing within a conduit link of SWMM is governed by the conservation of mass and momentum equations for gradually varied, unsteady flow (i.e., the Saint Venant flow equations). The SWMM user has a choice on the level of sophistication used to solve these equations. There are three routing methods in SWMM: steady flow routing and kinematic wave routing and dynamic wave routing. Each of these routing methods employs the Manning equation to relate flow rate to flow depth and bed (or friction) slope. For user-designated Force Main conduits, either the Hazen-Williams or Darcy-Weisbach equation can be used when pressurized flow occurs.

Surface pollutant are set to accumulate by both linear and nonlinear accumulation with time. Pollutants from different land uses are simulated using the wash-off approach. The wash-off is a simple function of runoff and as a first-order decay of available pollutant on the watershed surface. However, this model doesn't consider re-introduction of pollutants into the drainage system and also existing groundwater pollution is neglected [13]. The latest release of this model offers five distinct types of LID features that can be placed in the sub-areas as an alternative to traditional BMPs. Currently, the model is capable of simulating bio-retention cells also known as rain gardens, street planters, green roofs, infiltration trenches, porous pavements, rain barrels, and vegetation swales.

Flows and pollutants are routed through the sewer system using a modified kinematic wave approximation and assume complete mixing. Hydrographs along the sewer system are modified to represent storage and treatment effects in the system. This includes retention time, treatment efficiency as well as estimating costs. Pollutants in storage systems can be modeled as either complete mixing or plug flow. The stand-alone module EXTRAN permits SWMM to rout inflow hydrographs through an open channel and a closed conduit system using an explicit numerical solution of the shallow water wave equations.

Following are the principal applications of the SWMM model for urban watersheds:

- Mapping flood plains of natural channel systems – SWMM 5 is a FEMA-approved model for National Flood Insurance Program studies.
- Designing and sizing drainage system components, including detention facilities.
- Sizing detention facilities and their appurtenances for flood control and water quality protection.
- Evaluating the effectiveness of best management practices and low impact development for reducing wet weather pollutant loadings.
- Controlling site runoff using green infrastructure practices as low LID controls.

- Designing control strategies for minimizing combined sewer overflows.
- Evaluating the impact of inflow and infiltration in sanitary sewer overflows.
- generating non-point source pollutant loadings for TMDL studies.
- Sizing detention facilities and their appurtenances for flood control and water quality protection.

SWMM has been applied in watersheds ranging from 2 ha to 40000 km^2 [25]. However, it has been mainly applied to watersheds smaller than 2 km^2. Further reading and case studies of the SWMM model along with the link for the model download can be accessed at www.epa.gov/water-research/storm-water-management-model-swmm.

Hydrological Simulation Program--Fortran (HSPF)

Stanford Watershed Model of the 1960s was the origin of the HSPF model and in early 1970s water-quality processes were added. Development of a Fortran version incorporating several related models using software engineering design and development concepts was supported by the Athens, Ga., Research Lab of EPA in the late 1970s. The development of the pre-processing and post-processing software algorithm and use of the USGS Watershed Data Management (WDM) system was developed jointly by the USGS and EPA. In the 1990's USGS developed an interactive version of the Fortran program. The current release of the HSPF is Version 11.

The model contains hundreds of process algorithms developed from theory, laboratory experiments, and empirical relations from instrumented watersheds. It has the capability to run as a continuous or single event for simulation of runoff quantity and quality both in urban and rural areas. It is applicable to run non-point sources of water quality from urban and agricultural land uses.

Extensive data like meteorological records of precipitation and estimates of potential evapotranspiration are required for watershed simulation using HSPF model. Air temperature, dewpoint, wind, and solar radiation are also required for snowmelt simulation using HSPF. Air temperature, wind, solar radiation, humidity, cloud cover, tillage practices, point sources, and (or) pesticide applications may be required for water-quality simulation using this model. Physical measurements such as the land area, channels geometry, and reservoirs geometry are necessary parameters for the model simulation. Thus, though it is considered one of the most physically based and tested catchment models, it is very complex and difficult to use. GIS processing enabled the development of the Better Assessment Science Integrating point and Nonpoint Sources (BASINS) program that can be used to process most of the watershed data required for setting up the HSPF model. HSPF model is capable of simulating surface runoff, interflow, base flow, snowpack depth and water content, snowmelt, evapotranspiration, groundwater recharge, interception soil moisture in continuous periods of time. It can also be used to estimate nonpoint-source loads from various land uses fate and transport processes as well in streams and lakes.

HSPF model has been used in watersheds ranging from 3-km^2 to 70000 km^2 [25]. In a streamflow modeling study at an urbanized 14.8-km^2 headwater portion of the Stroubles Creek watershed in Montgomery County, Virginia, it is found that HSPF model was better than SWMM in predicting baseflow whereas SWMM predicted peak flow better than HSPF [25]. Besides, Global sensitivity analysis of this study found HSPF more sensitive to groundwater parameters and SWMM to imperviousness. Further reading and case studies of the HSPF model along with the link for model download can be found at the following address: www.epa.gov/ceam/hydrological-simulation-program-fortran-hspf.

Windows Technical Release-55 (TR-55)

The U.S. Department of Agriculture (USDA) Soil Conservation Service (NRC) which is now known as the Natural Resources Conservation Service (NRCS) developed the Technical Release 55 (TR-55; Urban Hydrology for

Small Watersheds) in 1975 as a simplified procedure to calculate single event storm runoff volume, peak rate of discharge, hydrographs and storage volumes in small urban watersheds. The model generates hydrographs from both urban and agricultural areas and at selected points along with the stream system. This program was originally developed for Windows DOS and in 1998, Technical Release 55 and the computer software were revised with big improvements in input and editing capabilities to what is now called WinTR-55. It uses the Win TR-20 program as the main engine for better hydrologic analysis. The changes in this revised version of TR-55 include upgraded source code to Visual Basic, which significantly changed the interface for the data entry, development of a Windows interface and output post-processor, enhanced capability for hydrograph-generation and flood routing through stream reaches and reservoirs.

WinTR-55 is a single event model with the input/output interface making easier to generate, route and add hydrographs. The model interpolates one hydrograph to the time interval of another one to add them together to determine peak flows. For example, if there are two hydrographs, Hydrograph A and Hydrograph B, of time intervals of 0.1 hours and 0.33 hours, respectively. The program will interpolate Hydrograph A to a time interval of 0.33 hours and add to Hydrograph B. Hydrograph B will be interpolated to an interval of 0.1 hours and add to Hydrograph A. Whichever hydrograph has the higher peak is saved for further operations downstream. Hydrographs are routed downstream through channels and/or reservoirs.

TR-55 uses the NRCS Curve Number (CN) approach for determining the watershed hydrology. The primary model input data includes sub-area acreage, CN (model determines based on the subarea land use description), rainfall distribution (model populates design storm distribution based on the project location), reach (stream) characteristics. While the model can also be used to design stormwater detention structures, it is furnished only with three types of outlet structures and short tube flow assumption potentially results in an inaccurate discharge estimate [26]. Thus, a different tool is recommended if a more accurate result is expected for structural design. Multiple sub-areas can be modeled within the watershed. A rainfall-runoff analysis can be performed on up to ten sub-areas and up to ten reaches.

However, the total drainage area cannot exceed 25 square miles. Additional information along with the model download options can be found at https://www.nrcs.usda.gov/wps/portal/nrcs/detailfull/national/water/?cid=stelprdb1042901.

MIKE URBAN

DHI developed the MIKE URBAN for analysis of water distribution systems, stormwater drainage systems and sewer collection in separate and combined systems. It is based on ESRI GeoDatabase for storing network as well as hydraulic. MIKE URBAN can be used for GIS-based model building and management, integrated water quality, fire flow, real-time control, and water distribution simulation, thematic mapping and integrated dynamic result visualization, integrating directly with the online and real-time control system. It's main module MIKE URBAN Model Manager incudes SWMM5 engine for the modeling of stormwater (i.e., surface runoff, flows, sediment transport), sewer modeling and US EPA EPANET 2.0 for the water distribution modeling like computation of steady-state, quasi-steady-state flow, pressure conditions and water quality in pipe networks. Important pipe flow features featured in the model are steady-state analysis, extended period analysis, and rule-based control. Water quality features used in this model are water age analysis, chlorine concentrations analysis, path and concentrations of pollutants analysis.

Runoff modeling using MIKE URBAN requires rainfall data, land use information, topographic map, soil type, green infrastructure and LID features as model input. The model furnishes results as time series, profile plots, animations, result comparisons, statistical analysis (min, max, accumulated values). In addition to hydrology, the model is also capable of providing water level in nodes, node flood, discharge in links, link depths, weir's an orifice's absolute discharge, etc. Mike URBAN has an advantage over SWMM due to its 2D simulation capability whereas SWMM is capable of 1D simulation only [27]. Further reading and case studies along with model download option can be found at https://www.mikepoweredbydhi.com/products/mike-urban.

The Hydrologic Modeling System (HEC-HMS)

HEC-HMS is a hydrologic modeling System designed to simulate watershed surface runoff and streamflow. This model simulates the physical properties of natural systems like river basins and hydrologic elements are connected in a dendritic network to simulate runoff processes. Meteorological data i.e., radiation, precipitation, snowmelt is required for simulating some of the simulations. This model was developed in 1992 by the Hydrologic Engineering Center (HEC) of the US Army Corps of Engineers in Davis, CA. The primary goal of developing HEC-HMS was to replace existing HEC-1, with additional capabilities such as the MODClark quasi-distributed (2D) hydrologic model. The HEC-HMS is modernized with advances in the numerical analysis that take advantage of the significantly faster computing power of recent time desktop machines. Besides, it has the features of grid cell surface hydrology and continuous simulation along with the graphical user interface that is not available in the HEC-1 model. Apart from those differences, most of the HEC-HMS model is quite similar to the predecessor HEC-1 which was considered an ideal model for hydrologic simulation. HEC-GeoHMS, a preprocessor for generating the features for the model describing the watershed is available from the HEC. HEC-HMS uses Modified Clark method to discretize the watershed domain into uniform grids, each of which acting as a quasi sub-area [28]. Though Modified Clark method distributes precipitation and infiltration parameters across the grids it lacks surface runoff transformation. As a result, variation in the amount of surface runoff will not be reflected in the HEC-HMS. In this case a physically distributed model: Gridded Surface and Subsurface Hydrologic Analysis (GSSHA) will be more appropriate as it incorporates a variation of precipitation, infiltration and surface runoff [28].

HEC-HMS model can be applied in a variety of geographic areas for solving a range of scenarios. It can be used for both small urban watershed runoff simulation to large river basin water supply design and flood hydrology. Watershed model using HEC-HMS is constructed by separating the water cycle into manageable pieces and constructing boundaries around the watershed of interest. Any mass or energy flux in the cycle can then be represented with a mathematical model. In most cases, several model choices are available for representing each flux. Each mathematical model included in the program is suitable in different environments and under different conditions. Making the correct choice requires knowledge of the watershed, the goals of the hydrologic study, and engineering judgment.

HEC-HMS model has been applied in different regions under different climatic conditions. It has been used as an event-based model as well as a continuous model for hydrological processes. The HEC-HMS model has been also used to simulate the rainfall-runoff process with geo-informatics and atmospheric models for flood forecasting and early warnings across the world [29-41]. The results from those researches are in good agreement with spatially and temporally predicting watershed response and is also demonstrated for simulating various scenarios in flood forecasting and early warnings [42]. Further details with model download options can be found at https://www.hec.usace.army.mil/software/hec-hms/features.aspx .

Urban Volume and Quality (UVQ)

UVQ provides the assessment of water supply, stormwater and wastewater services in urban areas in a conventional or unconventional way. UVQ is an effective preliminary assessment tool for determining the impacts of urban development options on the total water cycle, as well as the performance of a wide range of non-conventional demand and supply side management techniques [43].

As a successor of Aquacycle, it has a contaminant balance to the water balance, and an improvement in the user interface [43]. According to Last[44], the main strengths of UVQ include simplicity, rapid runtime and description of the cityscape. Weaknesses include the predominant focus on residential areas, range of indicator outputs, limited availability of water management techniques, and limited consideration of the natural systems.

Mitchell and Diaper [43] highlighted the fact that UVQ was developed "with the objective of maximum applicability to all urban areas in both Australia and Europe." As a result, it can model a variety of land use types (not just residential); a range of different conventional water infrastructure technologies and account for local climatic conditions. One disadvantage of UVQ is that it can only handle one climate file, which assumes a constant climate for the whole model [43].

UVQ divides the catchment into site/unit block, local/neighborhood, and regional/catchment [13, 43]. Input data characteristic can be specific to study site or generic depending on data availability and requirements. Land use categories can be urban/rural residential, commercial business, commercial horticulture, general light industry, parks and gardens, and rural open space. Detailed information on the features such as indoor/outdoor water demands, garden, roads, lot size, paved areas, infiltration, exfiltration capacities, etc. are required. For water quality analysis, contaminant data for all streams (sources) are as concentrations.

An array of results including water flows and quality from different spatial units, neighborhoods, and from the whole catchment block and associated water reuse alternatives can be produced at daily, monthly and yearly time steps [43]. Besides, monthly or yearly contaminant balance values are can be generated for the different spatial scales. While more complex models with diverse capabilities are now available, UVQ is still seen as a viable model for the urban water hydrology analysis. Though this model is categorized as a complex model, relative to other complex models discussed in this section, it is simple to use and is also available for free download. Detailed information can be found at https://www.wateronline.com/doc/part-3-urban-volume-and-quality-a-water-and-c-0002.

Model for Urban Stormwater Improvement Conceptualisation (MUSIC)

MUSIC was initially introduced in March 2001 as a beta version shortly after which the second version was released with extended features and provided export options. Later, Version 3 of the MUSIC model introduced life-cycle costing, rainwater tanks, and infiltration basins. Version 4 provided advanced modeling of bioretention and infiltration systems, as well as simpler calibration tools and improved support. This model can be used both as a continuous or event-based model. It can take up to 100 km square area of the catchment with time steps varying from 6 min to 24 hours. The model has been improved with continued research of the data modeled data gathering and testing for years by the Cooperative Research Centre for Catchment Hydrology (CRCCH), Australia [45]. The latest version MUSIC Version 6 introduced a major new feature called MUSIC-link, which streamlines the process for assessing the compliance of Water Sensitive Urban Designs (WSUD) which check for compliance of submitted land development plans against guidelines from regulatory agencies. It generates efficiencies in the development process by not only providing very specific modeling parameters to designers but also speeding up the assessment process.

Data requirements for this model include rainfall and meteorological data, catchment characteristics and conceptual design of stormwater treatment measures [45]. Though this model is not a detailed design tool it can be used as a conceptual decision support medium. The model predictions are used to evaluate and compare alternative strategies to protect aquatic environments from land use changes. MUSIC can model a wide range of treatment controls to find the best way to capture and reuse stormwater runoff, control water pollution, and reduce the frequency of runoff. It addresses contaminant removal approximately through the buffer strip removal equation. Bio-retention devices or filtration devices included in this model is limited to water quantity aspects, however, it addresses water quality, infiltration, storage, and hydraulic aspects using vegetative swales. The algorithm for routing contaminants through a channel reach in MUSIC is based on continuity of contaminant mass equation [45]. The disadvantage of the MUSIC model is that it has the capability to consider only event-mean contaminant concentration but not continuous pollutograph. More information on MUSIC can be found at http://urbanwater-eco.services/project/model-for-urban-stormwater-improvement-conceptualisation-music/.

Storage, Treatment, Overflow, Runoff Model (STORM)

Water Resources Engineers, Inc. (WRE) of Walnut Creek, California developed the STORM model in the early 1970s as part of a project for HEC of the U.S. Army Corps of Engineers. It is a Fortran based analysis tool for estimating runoff quantity and quality from urban and rural watersheds [46]. Further development of the model was carried out by both the HEC and Resource Analysis Inc. of Cambridge, Massachusetts. However, the emergence of other more advanced models has decreased its popularity in recent years.

It can be used both as a continuous and event-based model which considers the interaction of snowmelt, dry weather flow, pollutant accumulation, and wash-off, land surface erosion, treatment processes, rainfall-runoff, reservoir storages, etc. This lumped conceptual model is mainly adopted for use in modeling processes of stormwater flow such as runoff generation, collection, storage, treatment, and discharge to receiving waters [47]. However, it is also useful for sizing of storage and treatment facilities in urban areas. Hourly precipitation, land use data including runoff parameters, pollutant accumulation, wash-off data, and land surface erosion data will be required for STORM model execution [47]. STORM models hydrology of a watershed as the accumulation of runoff from upstream sub-areas [18]. There are three methods for calculating hourly runoff in STORM model: coefficient method, soil-complex cover method, and the unit hydrograph method.

Oversimplified runoff calculations of this model based on simple linear relationships commonly results in inaccurate predictions. It calculates pollutant accumulation as proportional to time and wash-off which is (wash-off) proportional to the amount of pollutant remaining. However, the lack of model calibration features and oversimplification of the routing procedures makes this model less favorable for stormwater management studies.

Source Loading and Management Model (WinSLAMM)

WinSLAMM is a non-point source stormwater quality model developed by John Voorhees and Robert Pitt for evaluation of nonpoint pollution in urban areas. The program was originally developed as outgrowths of early stormwater research conducted during the 1970s for the US EPA and other agencies. The model is based on field observations of land use specific pollutants concentration and performance characteristics of grass swales, wet detention ponds, and street sweeping, as either source area or outfall control practices. The focus of the model is on small storm hydrology and particulate wash-off. The main advantage of WinSLAMM model is that its evaluation capability at the source area level allows stormwater quality professionals the ability to target the highest loading areas and recommend improvements to reduce runoff volume and pollution loading from those areas.

Unlike many stormwater models, it does not lump impervious areas together nor does it lump all the areas in single land use together. The WinSLAMM model may be obtained from PV & Associates, LLC (www.winslamm.com/default.html). The model provides an extensive set of rainfall, runoff, and particulate solids and other pollutant files developed from the National Stormwater Quality Data Base for most urban areas in the county. WinSLAMM is a highly regarded model for evaluating LID practices and GI applications for non-point source control. However, this model does not include flow routing and pollutant fate and transport analysis in conveyance systems.

The most recent development of the WinSLAMM model includes the interface of WinSLAMM model with ArcGIS. There are two versions of GIS interfaced WinSLAMM models, namely, ArcSLAMM and ArcSLAMM Plus. The graphical interface allows users to define both source area and drainage system stormwater control practices using a drag-and-drop interface, and the program and web site provides extensive program help and stormwater quality references. As detailed by the developers of the model, while ArcSLAMM is used for smaller watershed modeling, ArcSLAMM Plus is used for large watersheds or municipality-scale studies using generalized land use information. The ArcSLAMM Plus tools generate a WinSLAMM standard land use model for every unique watershed, land use, and soil type combination in the study area to easily generate existing pollutant loads for a large project site. The tool can also be used to create graphics displaying spatially variable unit pollutant loads to help with planning efforts. Detailed information about this model can be found at http://www.winslamm.com.

System for Urban Stormwater Treatment and Analysis Integration (SUSTAIN)

SUSTAIN was developed by the USEPA primarily to assist stormwater decision making by developing and implementing plans for stormwater flow and pollutant controls on a watershed scale. It performs hydrologic and water quality modeling in watersheds and urban streams. SUSTAIN contains seven modules that integrate with ArcGIS. However, if the GIS interface is not required, SUSTAIN can also be used without GIS. Hydrology, hydraulics, and pollutant loading are computed using EPA SWMM, Version 5. Sediment transport is based on HSPF. The available modules are Framework manager for controlling data exchange between system components, BMP siting tool for selecting suitable locations for BMPs based on drainage area, hydrological soil group, groundwater depth, buffer etc.

Land simulation module for importing externally generated time-series data and computing runoff and pollutant loads from land based on either SWMM (hydrograph and pollutograph) or HSPF algorithm (sediment transport), BMP simulation module to simulate flow and transport phenomena like flow routing, infiltration, evapotranspiration, pollutant routing and removal, buffer strip flow routing, strip sediment trapping, and pollutant removal for BMPs, Conveyance simulation module performs routing of flow and pollutants through a conduit using SWMM or HSPF algorithms, BMP optimization module for identifying cost-effective BMP placement and selection strategies based on a pre-determined list of feasible sites and applicable BMP types and size ranges, and a Post-processor for analyzing and displaying the simulation outputs at multiple locations, and for scenarios and parameters of interest.

The data needs for SUSTAIN model is very similar to other complex watershed models. It generally includes Geodatabase file with spatial and tabular data, GIS layers and lookup tables information, suitability criteria(optional) and locations of BMP, assessment points, cost database file. Most models that operate on an hourly or shorter time step, such as HSPF, SWMM, and WINSLAMM, are compatible with SUSTAIN. When importing information from an externally generated model, the SUSTAIN application builds on the documentation, testing, calibration, and validation of the external model.

In SUSTAIN, the optimization module synthesizes information from the BMP, land, and conveyance modules and generates solutions that are looped back for evaluation using the same modules again. Via this evolutionary search process, the optimizer identifies the best or most cost-effective BMP solutions according to the user's specific conditions and objectives. Finally, the post-processor analyzes optimization results using specific graphical and tabular reports that facilitate the classification of storm events for analysis, viewing the time series of specific storm events, evaluating BMP performance by a storm event, and developing the cost-effectiveness curves for treatment alternatives.

Further details can be found at https://www.epa.gov/water-research/system-urban-stormwater-treatment-and-analysis-integration-sustain.

Gridded Surface Subsurface Hydrologic Analysis (GSSHA)

The US Army Engineering Research and Development Center's multidimensional modeling tool GSSHA is a significant enhancement of FORTRAN based CASC2D model. Dr. Bahram Saghafian at the U.S. Army Construction Engineering Research Laboratory transformed the FORTRAN code to C programming language with enhancement and reformulation. GSSHA is gaining increasing interest and attention among the research and engineering community because of its capability to perform complex studies of atmospheric, land-based, wetland, riverine, and coastal systems to help predict and mitigate watershed management. It is a 2D hydrologic model that divides the watershed into square grids of equal size and uses physical equations to simulate the hydrologic processes like floodplain analysis of single or continuous events and represents the watershed as-close-as-possible. Watershed Management System (WMS) provides a custom interface with pre and post user-friendly processors to run and visualize the results for GSSHA model.

Geospatial data required for the GSSHA model can be extracted and set easily using WMS. It is capable of groundwater modeling (i.e., exfiltration and discharge), contaminant transport, sediment transport, snowmelt, storm and tile drain modeling.

Manning's equation is used in GSSHA model similar to SWMM to calculate the overland flow from the overland depth. However, GSSHA solves it in 2D whereas SWMM solves in the 1D scheme. To calculate infiltration, GSSHA uses the Green-Ampt or Richard's equations. It can also calculate infiltration using multilayered Green and Ampt & Green and Ampt with redistribution method. GSSHA employs cell by cell discretization in the channels providing an advantage over SWMM model which uses uniform size for channels between the nodes [48].

Unlike HEC-HMS which calculates hydrology based on sub-watershed average conditions, GSSHA calculates hydrological processes on a cell by cell basis providing more spatial detailed calculations. Therefore, cell-to-cell rainfall, surface runoff calculations of GSSHA require longer computation time than HEC-HMS and SWMM models [49].

GSSHA model can track the fate of associated pollutants through the coupled system and can provide soil moisture, runoff, and flooding predictions that can be used to asses fire threat, irrigation needs, and utilize boundary conditions to simulate coastal flooding. Data requirement of the model increases when the model becomes comprehensive for simulating complex scenarios. For a simplistic rainfall-runoff simulation model, GSSHA requires at least the Green-Ampt parameters, surface roughness (overland roughness), channel roughness, and initial soil moisture at the grid cell level. Besides that, the topography and watershed information is also necessary [28].

GSSHA also includes the ability to simulate time and space varying head boundary conditions, making it suitable in many coastal watershed applications. GSSHA was used to predict the inland effects of predicted storm surge from Hurricane Irene on New York City and Long Island. More information about GSSHA can be found at https://gsshawiki.com/Gridded_Surface_Subsurface_Hydrologic_Analysis.

Watershed Management System (WMS)

WMS is a watershed management platform that offers state-of-the-art tools to perform automated watershed delineation and hydrologic modeling with automatic computation of geometric basin data such as area, mean elevation, maximum flow.

It is used to set up and run hydrologic models. WMS supports most of the industry-standard hydrologic models like HEC-HMS, HEC-RAS, TR-55, HSPF, EPA-SWMM, etc.

It serves as an all-in-one watershed package with a graphical user interface, pre and post processors, web-based data acquisition, terrain data import & editing tools, and step by step modeling wizards for several watershed models. One of the most distinguishing advantages of WMS is its built-in GIS environment and capability to display and overlay world coordinates.

It can compare results of a watershed developed in two different models that are supported by WMS thus helps in decision making. Another major advantage of WMS over other popular models is its ability to manipulate digital terrain data for hydrologic modeling because of it's built-in GIS environment [28]. It can export and import hydrologic model data from EPA SWMM to WMS and vice versa. It comes with a complete SWMM interface and a user-friendly and fully functional interface for the USCE's distribute GSSHA model [28].

WMS also provides a hydraulic interface making it compatible with HEC-RAS. The RAS model can be run as a steady or unsteady state, and results are used to delineate floodplain extents and animations of flood waves for complete flood plain analysis. It has an integrated hydraulic toolbox and incorporates the latest release of the widely used Federal Highways culvert design model, HY-8 to delineate a watershed upstream from a culvert, compute hydrographs, determine inundated area behind a culvert and analyze culverts [28]. Further details about WMS can be found at https://www.aquaveo.com/software/wms-watershed-modeling-system introduction.

Table 1 presents a list of most commonly used urban watershed models along with their source and functionality. The information in Table 1 can assist users to select the appropriate model to meet their goals considering the availability of the model input parameters data and other resources. In general, it is suggested to start experimenting with the simpler models that require relatively fewer input parameters data and resources and advance into more complex models depending on the expertise and availability of watershed-specific data.

New Generation Urban Watershed Models

Machine learning or deep learning has become a buzz word in today's world. Initiating artificial intelligence (AI) using a machine learning algorithm is dated back to 1950s. However, this idea was not as popular as today due to the limitation of computing power and data availability.

Table 1. Selected urban watershed models with source and functionality information

Model	Developer	Functionality			Resolution		Hydrology		Water quality			Accessibility		Model type
		Planning	operational	Design	Event based	Continuous	Surface runoff	Subsurface runoff	Sediment	Nutrient	Pesticides	Public	Commercial	
HEC-HMS	US Army Corps of Engineers		*		*	*	*		*	*			*	1D/2D
SWMM	USEPA	*		*	*	*	*	*	*	*		*		1D
XP SWMM	XP Solutions	*	*	*	*	*							*	1D/2D
MIKE URBAN	DHI Water & Environment	*	*	*	*		*	*	*	*			*	2D
HSPF	USEPA	*		*	*	*	*	*	*	*	*	*		1D
DR3M	US Geological Survey	*		*	*		*	*				*		1D
Penn State	Penn State University				*	*						*		1D
STORM	US Army Corps of Engineers	*			*	*	*	*	*			*		1D
Hydro Planner	CSIRO	*										*		1D
ILLUDAS	British Road Research Laboratory	*			*							*		1D
QQS	e-Water	*		*	*	*	*	?				*	*	1D
MUSIC	CRCCH	*		*		*			*	*			*	1D
GSSHA	USACE	*		*	*	*	*	*	*	*	*	*		1D/2D
WMS	Aquaveo	*		*	*	*	*	*	*	*	*		*	1D/2D

QQS = Quantity-quality Simulation; ILLUDAS = The Illinois Urban Drainage Area Simulator.
DR3M = Distributed Routing Rainfall-Runoff Model; XP SWMM = Storm Water Management Model by XP solutions;

Now, data processing power is increasing day by day with higher processing speed, bigger ram, and storage capacity. This power has given immense ability to see the unseen "intelligence" from the large quantities of data also known as "big data." Our world and its problems are full of patterns. In machine learning, the algorithm (software) can learn patterns with experience or data pass through the model and thus make improvement in the output or prediction from time to time. This pattern learning is a training process and is similar to teaching a child by showing him/her different shapes of the same alphabet. In this way, once the algorithm learns patterns of any system, it can easily predict what is going to happen in the future.

One of the most powerful algorithms of AI is Artificial Neural Network (ANN) where the model (algorithm) is developed in many layers connecting with each other mimicking the neurons of the human brain. Over the past decade, there has been widespread interest in the field of artificial intelligence and their application in engineering problem solving [50, 51]. Like many other engineering fields, modern-day water resources systems are equipped with hundreds of sensors and data-driven technologies collecting information in stages of the water source to supply end to smartly manage the whole process. This collected data setting works as a basis for big data analysis that helps identify and solve problems in water resources management. Besides, these sensors can even share water data over the internet on a real-time basis to and with an AI program which can optimize and improve the water resources system facilities in real-time. AI models working with big water data can provide insights to make the water resources system more sustainable by predicting the changes, both quantity, and quality, and prevent or minimize possible negative impact before it happens. The main purpose of the AI model is to replace the physical models; however, it may induce more uncertainties into the system.

At present, many hydrodynamic and hydrologic models for water resources system are available and the modeling techniques have gradually improved over time. Multiple Linear Regression (MLR), Auto-Regressive Integrated Moving Average (ARIMA) techniques are classical linear black-box models (we can also call them statistical model) handling time series data have been used for years for forecasting and control of hydrological phenomena [52-54]. However, unpredictable climatic condition and other nonlinear and non-stationary phenomena pose a challenge to these models to accurately predict different hydrological and hydraulic process [55]. To overcome this situation, Artificial Neural Networks (ANN), the Decision Tree (DT), and Support Vector Machine or Regression (SVM or SVR) types of data-driven approaches are getting much popularity in solving different real-life complex problems due to their approximation capability of nonlinear functions [56-58]. For solving hydrological problems like streamflow prediction and control least square support vector regression (LSSVR), multivariate adaptive regression splines (MARS) and M5 Model Tree (M5-Tree) approaches are widely used. LSSVR is the enhanced modified version of SVR which is used to predict evapotranspiration [59], water demand modeling [60], river water pollution forecasting[61]. ANN is also proved useful to evaluate water quality objectively and quickly on the basis of continuous water sampling and monitoring [48], flood forecasting [49, 50], rainfall-runoff forecasting [51-54]. Adaptive network fuzzy interference system (ANFIS) is an AI approach where a back-propagation neural network (BPNN) is adopted to predict reservoir water shortage rate and optimize reservoir operation [58]. Thus ANFIS has superiority over ANNs because it adopts fuzzy inference system with the ANN approach [62].

Backpropagation neural network (BPNN), a supervised learning network in ANN is good at nonlinear approximation and can be used to establish the water quality evaluation system using MATLAB programming toolbox. The central theme of BPNN algorithm is adjusting the weight of the network to the smallest total error. Algorithm of the BPNN model learning process consists of positive propagation and backpropagation. Through repeated positive propagation and backpropagation, neurons weights and thresholds of layers are modified continually to reduce the error function till the error function is no longer reducing. More information on BPNN and ANFIS can be found in Rumelhart et al. [63, 64].

In spite of ANN and ANFIS models good approximation of nonlinear problem, they have problems dealing with non-stationary statistical data whose mean (sometimes variance) changes with time. To solve this problem some hybrid models like chaotic neural networks [65], cluster neural networks [66], bootstrapped neural networks [66] principle component analysis (PCA) [67, 68], wavelet-based artificial neural network and wavelet-based adaptive neuro-fuzzy inference system [69], etc. are developed. Further study is suggested to choose the best algorithm for complex time series data analysis in order to forecast the water resources systems accurately.

CONCLUSION

Due to rapid urbanization and limited natural resources, different strategies are required for a sustainable hydraulic, hydrologic and water quality management in the urban areas. Once used conventional methods of water resource management are no longer preferred. Rise of computing power with the availability of personal computer propelled the urge to create new computation methods in hydrological science and computer modeling. Though lumped and semi distribute watershed models (i.e., lumped: Win TR-55, SWMM; semi-distributed: HEC-HMS) are used more widely than the distributed models due to their simple model setup and less computation time, newer distributed conceptual/physical computer models like GSSHA represent hydrological processes much more accurately than the earlier approximated models. However, semi-distributed models are often a better choice which can meet many of urban management needs if enough spatial input data is available. Addition of sustainability features, such as green infrastructure for urban stormwater management, are becoming more common in the recent model updates. Groundwater component and flow routing features for green infrastructure components are often simplified in computer models. General improvements such as increased spatial and temporal resolutions, better flow routing and groundwater components, including a wide range of possible water quality parameters, integrated flow, and water quality approach, the capability to assess green infrastructure controls, improved calibration technique and implementation of artificial intelligence are areas of attention for the model developers. Finally, in coming years, with rapid development in computer technology, urban watershed models are expected to be even more graphics intensive, user-friendly and more accurate in predicting watershed runoff water quantity and quality.

REFERENCES

[1] Hester, E. T. and K. S. Bauman, Stream and retention pond thermal response to heated summer runoff from urban impervious surfaces 1. JAWRA *Journal of the American Water Resources Association*, 2013. **49**(2): p. 328-342.

[2] Liu, A., et al., Influence of rainfall and catchment characteristics on urban stormwater quality. *Science of the total environment*, 2013. **444**: p. 255-262.

[3] Chen, J., et al., Urbanization impacts on surface runoff of the contiguous United States. *Journal of Environmental Management*, 2017. **100**(187): p. 470-481.

[4] Rosburg, T. T., P. A. Nelson, and B. P. Bledsoe, Effects of urbanization on flow duration and stream flashiness: a case study of Puget Sound streams, western Washington, USA. *JAWRA Journal of the American Water Resources Association*, 2017. **53**(2): p. 493-507.

[5] Lacher, I. L., et al., Scale-dependent impacts of urban and agricultural land use on nutrients, sediment, and runoff. *Science of the Total Environment,* 2019. **652**: p. 611-622.

[6] Daghighi, A., *Harmful algae bloom prediction model for western lake Erie using stepwise multiple regression and genetic programming.* 2017.

[7] Liu, J., Z. Shen, and L. Chen, Assessing how spatial variations of land use pattern affect water quality across a typical urbanized watershed in Beijing, China. *Landscape and Urban Planning*, 2018. **176**: p. 51-63.

[8] Whipple, W., et al., *Stormwater management in urbanizing areas. 1983*, Englewood Cliffs, New Jersey: Prentice-Hall, Inc. 234 pp.

[9] Burian, S., et al., *Historical Development of Wet-Weather Flow Management*. Vol. 125. 1999.

[10] Fletcher, T. D., H. Andrieu, and P. Hamel, Understanding, management and modelling of urban hydrology and its consequences for receiving waters: A state of the art. *Advances in Water Resources,* 2013. **51**: p. 261-279.

[11] Rangari, V., A. Kumar Patel, and U. Nanduri, *Review of urban stormwater models*. 2018.

[12] Obropta, C. and J. Kardos, *Review of Urban Stormwater Quality Models: Deterministic, Stochastic, and Hybrid Approaches* 1. Vol. 43. 2007. 1508-1523.

[13] Elliott, A. and S. A. *Trowsdale, A Review of Models of Low Impact Urban Stormwater Drainage*. Vol. 22. 2007. 394-405.

[14] Chow, V. T., D. R. Maidment, and L. W. Mays, *Applied hydrology. McGraw-Hill series in water resources and environmental engineering*. 1988, New York: McGraw-Hill.

[15] Dingman, S. L., *Physical hydrology*. 2nd ed. ed. 2002, Upper Saddle River, N. J.: Prentice Hall.

[16] Chow, V. T., *Handbook of Applied Hydrology*. 1964: McGraw-Hill Book Company, USA.

[17] Li, S. G. and D. McLaughlin, *A nonstationary spectral method* for solving stochastic groundwater problems: Unconditional analysis. 1991. **27**(7): p. 1589-1605.

[18] Zoppou, C., Review of urban storm water models. *Environmental Modelling & Software*, 2001. **16**(3): p. 195-231.

[19] Pitt, R. and G. A. Burton Jr, *Stormwater effects handbook: A toolbox for watershed managers, scientists, and engineers*. 2001: CRC Press.

[20] Zhou, Q., A *Review of Sustainable Urban Drainage Systems Considering the Climate Change and Urbanization Impacts*. Vol. 6. 2014. 976-992.

[21] Nix, S. J., *Urban stormwater modeling and simulation*. 1994: CRC Press.

[22] Wang, Z. M., O. Batelaan, and F. De Smedt, A distributed model for water and energy transfer between soil, plants and atmosphere

(WetSpa). *Physics and Chemistry of the Earth*, 1996. **21**(3): p. 189-193.

[23] Jajarmizad, M., S. Harun, and M. Salarpour, *A Review on Theoretical Consideration and Types of Models in Hydrology*. Vol. 5. 2012. 249-261.

[24] James, W., L. A. Rossman, and W. R. C. James. *User's guide to SWMM* 5:[based on original USEPA SWMM documentation]. 2010. CHI.

[25] Nayeb Yazdi, M., et al., An evaluation of HSPF and SWMM for simulating streamflow regimes in an urban watershed. *Environmental Modelling & Software*, 2019. **118**: p. 211-225.

[26] Pitt, R., *WinTR55 for Watershed Analyses*. 2005.

[27] Bisht, D. S., et al., Modeling urban floods and drainage using SWMM and MIKE URBAN: a case study. *Natural Hazards*, 2016. **84**(2): p. 749-776.

[28] Paudel, M., An examination of distributed hydrologic modeling methods as compared with traditional lumped parameter approaches. 2010.

[29] Al-Abed, N., F. Abdulla, and A. Abu Khyarah, GIS-hydrological models for managing water resources in the Zarqa River basin. *Environmental Geology*, 2005. **47**(3): p. 405-411.

[30] Rueedi, J., et al., Effect of different water management strategies on water and contaminant fluxes in Doncaster, United Kingdom. *Water Sci Technol,* 2005. **52**(9): p. 115-23.

[31] Anderson, M. L., et al., Coupling HEC-HMS with Atmospheric Models for Prediction of Watershed Runoff. *Journal of Hydrologic Engineering,* 2002. **7**(4): p. 312-318.

[32] Rathod, P., K. Borse, and V. Manekar, *Simulation of Rainfall -Runoff Process Using HEC-HMS*. 2015.

[33] Saleh, A., R. Ghobad, and N. Rostami, *Evaluation of HEC-HMS methods in surface runoff simulation* (Case study: Kan watershed, Iran). Vol. 5. 2011. 1316-1321.

[34] Emerson Clay, H., C. Welty, and G. Traver Robert, Watershed-Scale Evaluation of a System of Storm Water Detention Basins. *Journal of Hydrologic Engineering,* 2005. **10**(3): p. 237-242.

[35] H. Hu, H., et al., *Gridded Snowmelt and Rainfall-Runoff CWMS Hydrologic Modeling of the Red River of the North Basin*. Vol. March/April 2006. 2006.

[36] Knebl, M. R., et al., Regional scale flood modeling using NEXRAD rainfall, GIS, and HEC-HMS/RAS: a case study for the San Antonio River Basin Summer 2002 storm event. *Journal of Environmental Management,* 2005. **75**(4): p. 325-336.

[37] McColl, C. and G. Aggett, Land-use forecasting and hydrologic model integration for improved land-use decision support. *J Environ Manage,* 2007. **84**(4): p. 494-512.

[38] Yusop, Z., C. H. Chan, and A. Katimon, Runoff characteristics and application of HEC-HMS for modelling stormflow hydrograph in an oil palm catchment. *Water Sci Technol*, 2007. **56**(8): p. 41-8.

[39] K Yener, M., Ü. Şorman, and A. Sensoy, Modeling studies with HEC-HMS and runoff scenarios in Yuvacik Basin, Turkiye. 2003.

[40] Sahour, H., A. Mokhtari, and S. Shahabi Ghahfarokhi, *Rainfall-runoff modeling using remotely sensed data and the Hydrologic Modeling System* (HEC-HMS). Vol. 22. 2016. 163-173.

[41] Bhavsar, P. N. and J. N. Patel, Event-based rainfall–run-off modeling and uncertainty analysis for lower Tapi Basin, India. *ISH Journal of Hydraulic Engineering*, 2018: p. 1-10.

[42] Kishor Choudhari, B. P., Jagadish Chandra Paul, Simulation of rainfall-runoff process using HEC-HMS model for Balijorenala watershed, Odisha, India. *International Journal of Geomatics and Geosciences.* Volume 5, No 2, 2014.

[43] Mitchell, V. and C. Diaper, *UVQ: A tool for assessing the water and contaminant balance impacts of urban development scenarios*. Vol. 52. 2005. 91-8.

[44] Mackay, R. and E. Last, SWITCH city water balance: a scoping model for integrated urban water management. *Reviews in Environmental Science and Bio/Technology*, 2010. **9**(4): p. 291-296.

[45] Wong, T. H. F. F., Timothy David; Duncan, Hugh P; Coleman, John R; Jenkins, Graham A. A model for urban stormwater improvement conceptualisation. In *International Conference on Urban Drainage 2002. 2002*. Reston VA USA: American Society of Civil Engineers.

[46] Hydrologic Engineering, C., Storage, treatment, overflow, runoff model "STORM": user's manual. 1977, Davis, CA*: Hydrologic Engineering Center*, Corps of Engineers, U.S. Army. 1 v. (various pagings).

[47] Chin, H. C. W. R. L. F. o. E. U., *A Review of Stormwater Quality Models*. April 2006. 2006, University of New South Wales - Water Research Laboratory.

[48] Moore Mitchell, F., G. Vasconcelos Jose, and C. Zech Wesley, Modeling Highway Stormwater Runoff and Groundwater Table Variations with SWMM and GSSHA. *Journal of Hydrologic Engineering,* 2017. **22**(8): p. 04017025.

[49] ElHassan, A., et al., Performance of a conceptual and physically based model in simulating the response of a semi-urbanized watershed in San Antonio, Texas. *Hydrological Processes*, 2013. **27**: p. 3394-3408.

[50] Null, n., Artificial Neural Networks in Hydrology. I: Preliminary Concepts. *Journal of Hydrologic Engineering*, 2000. **5**(2): p. 115-123.

[51] Tang, H., et al., *Artificial intelligence model for water resources management*. Vol. 163. 2010. 175-187.

[52] Box, G. E., et al., *Time series analysis: forecasting and control*. 2015: John Wiley & Sons.

[53] Yaseen, Z. M., et al., Artificial intelligence based models for stream-flow forecasting: 2000–2015. *Journal of Hydrology*, 2015. **530**: p. 829-844.

[54] Abrahart, R. J., et al., Two decades of anarchy? Emerging themes and outstanding challenges for neural network river forecasting. *Progress in Physical Geography*, 2012. **36**(4): p. 480-513.

[55] Yaseen, Z. M., O. Kisi, and V. Demir, Enhancing Long-Term Streamflow Forecasting and Predicting using Periodicity Data Component: Application of Artificial Intelligence. *Water Resources Management,* 2016. **30**(12): p. 4125-4151.

[56] Hasebe, M. and Y. Nagayama, Reservoir operation using the neural network and fuzzy systems for dam control and operation support. *Advances in Engineering Software*, 2002. **33**(5): p. 245-260.

[57] Nourani, V., R. G. Ejlali, and M. T. Alami, Spatiotemporal groundwater level forecasting in coastal aquifers by hybrid artificial neural network-geostatistics model: a case study. *Environmental Engineering Science*, 2011. **28**(3): p. 217-228.

[58] Chang, F. J., Y. C. Wang, and W.-P. Tsai, Modelling Intelligent Water Resources Allocation for Multi-users. *Water Resources Management,* 2016. **30**(4): p. 1395-1413.

[59] Guo, X., X. Sun, and J. Ma, Prediction of daily crop reference evapotranspiration (ET0) values through a least-squares support vector machine model. *Hydrology Research*, 2011. **42**(4): p. 268-274.

[60] Hwang, S. H., D. H. Ham, and J. H. Kim, Forecasting performance of LS-SVM for nonlinear hydrological time series. *KSCE Journal of Civil Engineering*, 2012. **16**(5): p. 870-882.

[61] Kisi, O. and K. S. Parmar, Application of least square support vector machine and multivariate adaptive regression spline models in long term prediction of river water pollution. *Journal of Hydrology*, 2016. **534**: p. 104-112.

[62] Okkan, U., Using wavelet transform to improve generalization capability of feed forward neural networks in monthly runoff prediction. *Scientific Research and Essays*, 2012. **7**(17): p. 1690-1703.

[63] Rumelhart, D., G. Hinton, and R. Williams, Learning internal representations by error propagation, In *Parallel Distributed Processing,* DE Rumelhart, JL McClelland eds. 1986, MIT Press, Cambridge.

[64] Chang, F. J. and Y. T. Chang, Adaptive neuro-fuzzy inference system for prediction of water level in reservoir. *Advances in water resources*, 2006. **29**(1): p. 1-10.

[65] Karunasinghe, D. S. and S.-Y. Liong, Chaotic time series prediction with a global model: Artificial neural network. *Journal of Hydrology*, 2006. **323**(1-4): p. 92-105.

[66] Cigizoglu, H. K. and Ö. Kişi, Flow prediction by three back propagation techniques using k-fold partitioning of neural network training data. *Hydrology Research*, 2005. **36**(1): p. 49-64.

[67] Wang, W., et al., Forecasting daily streamflow using hybrid ANN models. *Journal of Hydrology*, 2006. **324**(1-4): p. 383-399.

[68] Wu, C., K. Chau, and Y. Li, Predicting monthly streamflow using data-driven models coupled with data-preprocessing techniques. *Water Resources Research*, 2009. **45**(8).

[69] Seo, Y., et al., Daily water level forecasting using wavelet decomposition and artificial intelligence techniques. *Journal of Hydrology*, 2015. **520**: p. 224-243.

In: Sustainable Water
Editors: V. Gnaneswar Gude et al.
ISBN: 978-1-53617-335-2

Chapter 4

NUMERICAL SIMULATION TO QUANTIFY PRESENT STATUS AND FUTURE PREDICTION OF WATER QUALITY OF TO-LICH RIVER, HANOI, VIETNAM

Pankaj Kumar*

Natural Resources and Ecosystem Services,
Institute for Global Environmental Strategies,
Hayama, Kanagawa, Japan
United Nations University,
Institute for the Advanced Study of Sustainability,
Tokyo, Japan

ABSTRACT

A few decades ago, the surface waterbodies in Hanoi, Vietnam, were an important source of water for various usages. However, because of the rapid global changes, the current status of most of the river system is very critical from the environmental, aesthetic and commercial usage points of

* Corresponding Author's Email: kumar@iges.or.jp.

view. Henceforth, this research work focused on assessing the current as well as predicting its future (2030) situation of four rivers (To-Lich, Set, Lu and Kim Nguu) using different scenarios while considering key drivers of global changes namely climate change and population growth along with the mitigation measures as mentioned in an existing master plan. Water Evaluation and Planning (WEAP), a predictive numerical simulation tool, was used to model river water quality using two scenarios viz., Business As Usual (BAU) and Scenario With Measures. The simulated output visualized future wastewater generation and indicated where additional appropriate countermeasures are needed in addition to the current master plan.

Results from this study show that current water quality for all streams is moderately to extremely polluted as compared with desirable category i.e., class B given by the local government. In addition, due to climate change and population changes, the water quality (expressed as BOD) will be further deteriorating by average 53.1% in 2030 as compared to the current situation. According to Hanoi city master plan for water resources, all domestic wastewater currently flowing into the rivers except for the Kim Nguu River will be transferred to new wastewater treatment plants. In such case, this study suggests that the rivers should be regularly supplemented with treated wastewater from surrounding wastewater treatment plants or water from other sources to keep the river flow and to improve the water quality significantly as shown in the mitigation plan scenario (85.8% reduction compared to the business-as-usual scenario).

Keywords: BOD, *Escherichia coli*, river pollution, wastewater management, water quality modelling, Hanoi

INTRODUCTION

Water is intrinsically linked to the sustainable and inclusive development of human societies. Yet, access to clean and safe drinking water is a major challenge (Heinemann et al., 2002; FAO 2016), which is well recognized under the global sustainable development goals (SDGs). To put things in perspective, more than 1.1 billion people do not have adequate access to clean drinking water and it is feared that nearly two-thirds of the nations will be under water stress by 2025 (Pink 2016; Kumar, 2018). In fact, researchers identified water insecurity as a major inhibitor of socio-economic growth in the developing world particularly in Asia and Africa, where increasing demand for water and contamination of existing water sources further complicate the scenario (Huizinga et al., 2017; Mukate et al., 2017). Although the concepts of water security vary within the existing scientific disclosures (Hoekstra et al., 2018), most researchers now agree that the water crisis in the 21st century is much more related to management rather than the actual scarcity of water (Tundisi, 2008). Of the various issues that impound water scarcity, one pressing issue is the poor quality of potable water and lack of water governance in most of the developing countries, which broadened the debates on water security over the last two decades. Consequently, the Sustainable Development Goals simultaneously focused on water quantity and quality compared to the erstwhile Millennium Development Goals, which only provided quantity related targets (Angelo et al., 2018). Nonetheless, a multitude of other factors, including, population growth, rapid urbanization and climate change are also expected to pose significant challenges in achieving future water security (Huizinga et al., 2017; Mukate et al., 2017).

Of the several strategies like wastewater infrastructure improvement, sewerage network improvement, nature based solution etc., discussed in the contemporary scientific disclosures, adaptive governance of water resources has been identified as an important tool for achieving water security. However, managing urban water environment is still a significant challenge (UN Water 2017; Ferguson et al., 2018); and the concept of urban water security is fundamentally different from the general water security concept (Hoekstra et al., 2018). A number of allied factors, including high population density, concentrated demand of fresh water, distribution system, recycling of wastewater and taxing are particularly important for an urban context. To meet such complexity, tapping reliable, all-weather sources remain the critical and first step for urban water planners. This has often led to tapping potential sources far away from the cities and resulted in massive investments in water infrastructure. Yet, renewable water resources have fallen dramatically on a per capita basis since the 1960s, particularly in South Asia. While it is estimated that about three billion people will reside in Asian cities by year 2050 (UNDESA 2015), countries like India, Pakistan and Afghanistan have already reached the water stress level in 2015, and several other countries like Nepal and Bangladesh are underway (Gareth et al., 2014). Thus, ensuring sustainable water supply and meeting the SDGs in growing Asian cities remains a critical challenge. Unfortunately, water resource planning in cities, particularly in developing nations has been done largely in a piecemeal manner, without taking into account of larger socio-economic factors (Downing 2012). While planners in developing countries are often obsessed with meeting the basic water demands through adequate supply, a critical step to achieve future urban water security is to integrate both hydrological and socio-economic factors. Understanding urban water security through a system perspective, including both the natural (i.e., source), social, economic and infrastructural components, therefore, remains highly imperative.

A number of holistic approaches have been conceptualized in the field of water resource management since the 1980's, of which, the integrated water resource management (IWRM) model has received the highest attention (Hoekstra et al., 2018). The IWRM model targets different components of water resource governance (socio-economic status, hydro-meteorological factors, agriculture, industries, wastewater etc.) and thereby, integrate them for science-led decision-making (Frija et al., 2015; Blanco-Gutiérrez et al., 2013). Several IWRM numerical models like RIBASIM (River Basin Simulation Model), MIKE, WEAP (Water Evaluation and Planning), WBalMo (Water Balance Model) have been developed and widely used to address water security issues. Yet, some of these models are data intensive, and therefore remain unusable for data-deficit regions. The WEAP model, nonetheless, is not data-intensive (Ingol-Blanco & McKinney 2013; Slaughter et al., 2014; Kumar et al., 2018), and thus, it was widely used in water quality modelling in developing countries. Moreover, the software package comes free of cost, to the benefit of water resource planners in developing countries.

Combination of high economic growth and uncoordinated rapid urban expansion results in unhealthy water environment especially around water bodies in most of the developing countries. Despite its importance, their current status and their management strategies for the near future is poorly documented. This study strives to assess the current situation and the simulated future status of water quality in the To-Lich River basin crossing Hanoi city, with the ultimate goal to help to formulate sustainable water resource management options for the area.

STUDY AREA AND METHODOLOGY

Study Area

Hanoi, the capital of Viet Nam, is located in the northwestern area of the Red River Delta. After the expansion of the administrative boundary in 2008, the area is 3,359 km^2, which is divided into 12 urban and 17 rural districts, and 1 district-level town (General Statistics Office, 2017). The climate is characterized as humid subtropical climate, with hot summer and rainy and mild winter. In addition to the three main rivers, namely the Red River, the Day River, and the Nhue River, there are many small rivers and canals consisting a dense water network, together with many ponds and lakes. The population of Hanoi was 7,202,900 as of 2015 (General Statistics Office, 2017), and it is projected to increase to 8.2 million in 2034 (General Statistics Office and United Nations Population Fund, 2016).

The study area is the watershed of the To Lich River system, consisted of four urban rivers flowing inside central Hanoi, namely the To Lich River, the Lu River, the Set River, and the Kim Nguu River (Figure 1). Due to its geographical characteristics, Hanoi, like many other Vietnamese cities, is prone to flooding. The city lies on the low and flat area, which contributes to the increased risk of floods in Hanoi. In addition, pollution of river water is also severe (Quan et al., 2010, Dao et al., 2010) due to low coverage of wastewater treatment.

Basic Information Regarding the Model and Data Requirement

The WEAP model was used to simulate future water quality variables in the year 2030 to assess alternative management policies in the To-Lich River basin. Here, according to the current master plan (VUWSW, 2015), water flowing in the targeted river (To-Lich) will be taken away to other place and treated water from Ho Tay Wastewater Treatment Plant (WWTP) will be diverted to To Lich River. For water quality modeling, wide range of input data including household discharge, past spatio-temporal water quality, wastewater treatment plants (Ministry of Natural Resources and Environment (MONRE)), population, historical rainfall, temperature (Vietnam Water Sanitation and Environment (VIWASE)), drainage networks (MONRE), river flow-stage-width relationships, river length, surface water inflows and land use/land cover (MONRE) is provided.

WEAP model was developed for the To Lich, Set, Lu and Kim Nguu River basin for four command areas with inter-basin transfers. Hydrologic modeling requires the entire study area to be split into smaller catchments with consideration of the confluence points and physiographic and climatic characteristics (Figure 1). The hydrology module within the WEAP tool enables modeling of the catchment runoff and pollutant transport processes into the river. Pollutant transport from a catchment accompanied by rainfall-runoff is enabled by ticking the water quality modeling option. Pollutants that accumulate on catchment surfaces during non-rainy days reach water bodies through surface runoff. The WEAP hydrology module computes catchment surface pollutants generated over time by multiplying the runoff volume and concentration or intensity for different types of land use.

During simulation, the land use information was broadly categorized into three categories, viz., agricultural, forest, and built-up areas. The soil data parameters were identified using previous secondary data and the literature. Result for the population distribution and its future trend at these four command areas were calculated by ratio method using VIWASE projection rate explained later in the chapter. Understanding the potential impacts of climate change is essential for informing both adaptation strategies and actions to avoid dangerous levels of climate change. In order to evaluate the climate change on water quality, this study has evaluated the future change in monthly average precipitation. Also, regarding future precipitation data, two different Global Climate Models (GCMs) i.e., MRI-CGCM3.2 and MIROC5; and Representative Concentration Pathway (RCP) i.e., RCP4.5 and RCP8.5 outputs were used after downscaling and bias correction. Statistical downscaling followed by trend analysis, a less computation demanding technique, which enables reduction of biases in the precipitation frequency and intensity (Goyal and Ojha 2011) was used here to get climate variables at monthly scale. Analysis of monthly precipitation data of 25 years period, both for historical (1980-2004) and future (2020-2044) was performed.

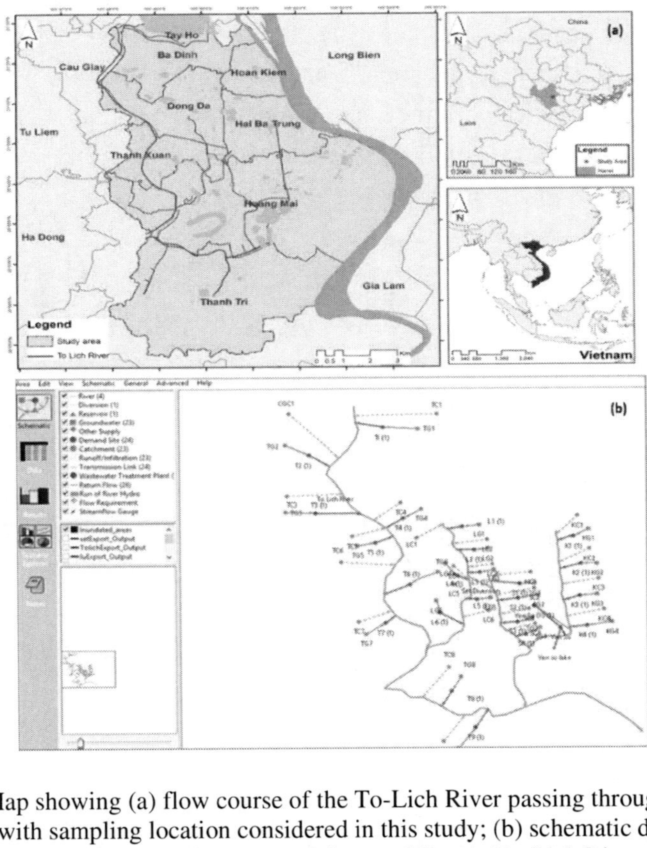

Figure 1. Map showing (a) flow course of the To-Lich River passing through Hanoi City along with sampling location considered in this study; (b) schematic diagram showing the problem domain for water quality modeling in To-Lich River using WEAP interface.

Result for the population distribution and its future trend at these four command areas were calculated by ratio method using UNDESA projected growth rate (UNDESA 2015).

Model Setup

The entire problem domain and its different components are divided into three catchments, which are further subdivided into twenty-three sub-basins considering influent locations of major tributaries (Figure 1). Other major considerations are the twenty-four demand sites and one wastewater treatment plant to represent the problem domain. Here, demand sites are meant to identify domestic (population) and industrial centers defined with their attributes explaining water consumption and wastewater pollution loads per capita, water supply source and wastewater return flow. Dynamic attributes are described as functions of time and include population and industries. Wastewater treatment plants are pollution handling facilities with design specifications that include total capacity and removal rates of pollutants.

The flow of wastewater into the To Lich, Set, Lu and Kim Nguu River and its tributaries mainly feeds through domestic, industrial and stormwater runoff routes. Here upflow anerobic sludge blanket reactor coupled with sequencing batch reactor (UASB-SBR) type of wastewater treatment plant is considered in the future water quality simulation as mentioned in the master plan and its treatment efficiency is assumed as 94% for COD, 97% for BOD, 77% for TN and 99.69% for fecal coliform (Khan et al., 2013). Also, there is no precise information available for the type and technology used for current WWTPs. So for the sake of simplicity, we have considered this advanced technology for both current and future WWTPs. In the absence of detailed information, the daily volume of domestic wastewater generation is estimated 130 liters of average daily consumption per capita based on literature review (UN WWAP, 2017).

For water quality, simulation is done using two possible scenarios as shown in Table 1. Simulation was done for the years 2015 and 2030 using 2011 as the reference year with consideration for population increase, climate change, wastewater generation and its treatment at wastewater treatment facilities.

First, business as usual scenario, where the effect of population growth and climate change using the average value of two GCM and two RCP on water quality of four different situations is observed, while keeping the capacity of all the existing wastewater treatment plants (200 MLD) constant by the year 2030.

For the scenario with measures, only To-Lich River was considered. Here, all the water in the river was diverted and all the treated water from Ho-Tay WWTP which lies north of the To-Lich River will be brought to To-Lich River.

Table 1. Summary of all the criteria considered for different scenarios in future water quality simulation

Scenario	Components
Business as usual	Climate change + population growth +WWTP of 200 MLD
With Measures	Climate change + population growth +Treated wastewater from Ho Tay WWTP will be transferred to To-Lich River

RESULT AND DISCUSSIONS

Precipitation Change

The comparative result of monthly precipitation pattern is shown in Figure 2. The observed annual precipitation for the year 2015 was 1644.5 mm, whereas the projected annual precipitation values for the year 2030 using MRICGCM3.2 GCM under RCP 4.5 and RCP 8.5 were 1637.4 mm and 1647.8 mm respectively. On the other hand, same projected precipitation value using MIROC5 GCM using RCP 4.5 and RCP 8.5 were 1658.9 mm and 1641.1 mm respectively. These values clearly indicates that the annual precipitation in the simulated from GCM output is not much different than the observed current value.

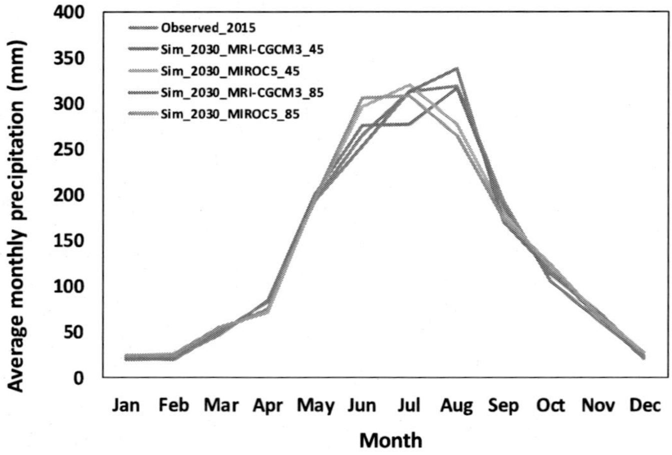

Figure 2. Graph showing a comparative study for current and future monthly rainfall at Lang *station.*

Population Growth

Future population is estimated from the ratio method using the data for Hanoi city from Vietnam Water, Sanitation and Environment (VIWASE, 2015) growth rate as shown in Table 2.

Table 2. Future projection of population growth in different catchment area using VIWASE projection rate

District	2015	2030	Area (km^2)
Ba Dinh	237,600	170,000	9.25
Cau Giay	242,500	147,500	12.03
Dong Da	387,800	255,000	9.96
Hoan Kiem	150,700	130,000	5.29
Hoang Mai	252,800	254,500	39.81
Tay Ho	143,300	119,800	24.01
Hai Ba Trung	311,200	200,700	10.09
Thanh Tri	211,600	355,488	62.93
Tu Liem	463,100	778,008	75.33
Thanh Xuan	244,100	135,000	9.08

Water Quality

Model Performance Evaluation

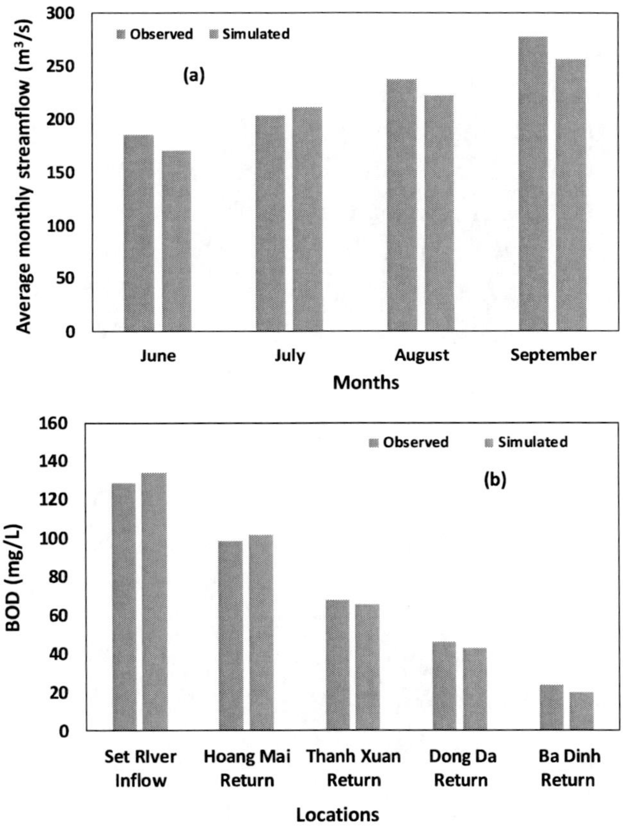

Figure 3. Validation of the model output by comparing simulated and observed (a) average monthly river discharge for the year 2013-2015 at Dong Da Return; (b) average BOD values for different locations for the year 2015.

Before doing future scenario analysis, the performance of the WEAP simulation was validated with observed and simulated values of hydrological and water quality parameters. Hydrology module parameters (effective precipitation and runoff/infiltration) were adjusted using trial and error method during simulation in order to reproduce the observed monthly stream flows for the period of 2013 to 2015 in case of hydrology module validation. The final best fitting values for both parameters were 92% and 55/45 respectively. Whereas, water quality simulation part was validated by comparing simulated and observed water quality concentrations at some observation points. Selection of above location and time was made on the basis of consistent availability of observed river discharge and water quality data respectively. Figure 3 (a) compares monthly simulated and observed stream flows at Dong Da Return (average value for the years, 2013-2015), showing that they largely match for most of the months with correlation coefficient (R^2) ≅ 0.82, root-mean-square error (RSME) ≅ 0.27, and an average error of 12%. The reason to select the three months for validation was that there is no water availability in the river especially during dry periods. Whereas, water quality simulation part was validated by comparing yearly average simulated and observed BOD concentration for the year 2015 at different location stretching from upstream to downstream. Selection of time i.e., year 2015 was made on the basis of consistent availability of observed water quality data. Results show a strong relationship between these two (Figure 3 (b)) (with the error of 15%) confirming suitability of the model performance in this problem domain.

Scenario Analyses

For water quality, simulation is done using two possible scenarios as shown in Table 2. Simulation was done for the years 2015 and 2030 using 2013 as reference year with consideration for population increase, land use change, wastewater generation and its treatment at wastewater treatment facilities.

Future simulation of water quality using selected parameters (BOD and *E. coli*) was conducted under two different scenarios (business-as-usual and with measures) were developed to project the river water quality in 2030. The simulation results for the water quality parameters using these two scenarios are shown in Figures 4, 5 and 6. Here small black bars on the simulated water quality in figure 4 and 5, shows the range because of change in GCM and RCP.

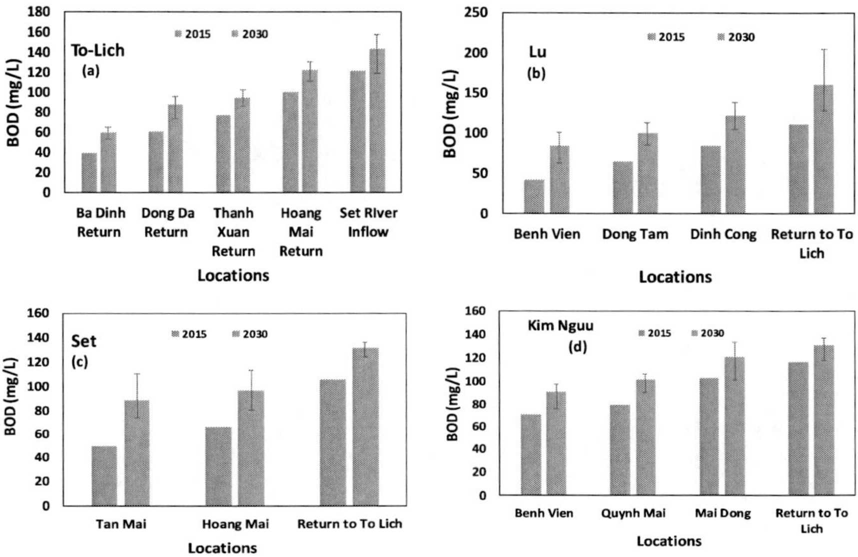

Figure 4. Results show simulated BOD values using business as usual scenario for (a) To-Lich, (b) Lu, (c) Set, and (d) Kim Nguu river systems.

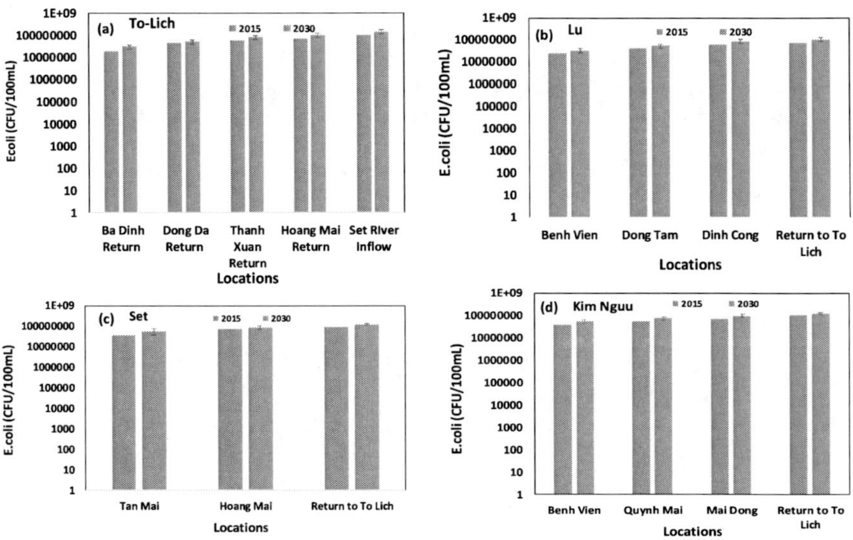

Figure 5. Results show simulated E.*coli*. values using business as usual scenario for (a) To-Lich, (b) Lu, (c) Set, and (d) Kim Nguu river systems.

Effect of both climate change and population changes are prominent in water quality status. It will deteriorate further in 2030 while comparing to the current situation. At present, water quality is worse in case of downstream sampling locations of most of the rivers. Looking into the water quality guidelines, many of the locations especially along the downstream do not comply for Class B for agriculture and aquaculture (< 25 mg/L) (Ministry of Natural Resource and Environment, 2015). Due to climate change and population changes, the water quality (expressed as BOD) will be further deteriorating by 53.1% on average in 2030 compared to the current situation.

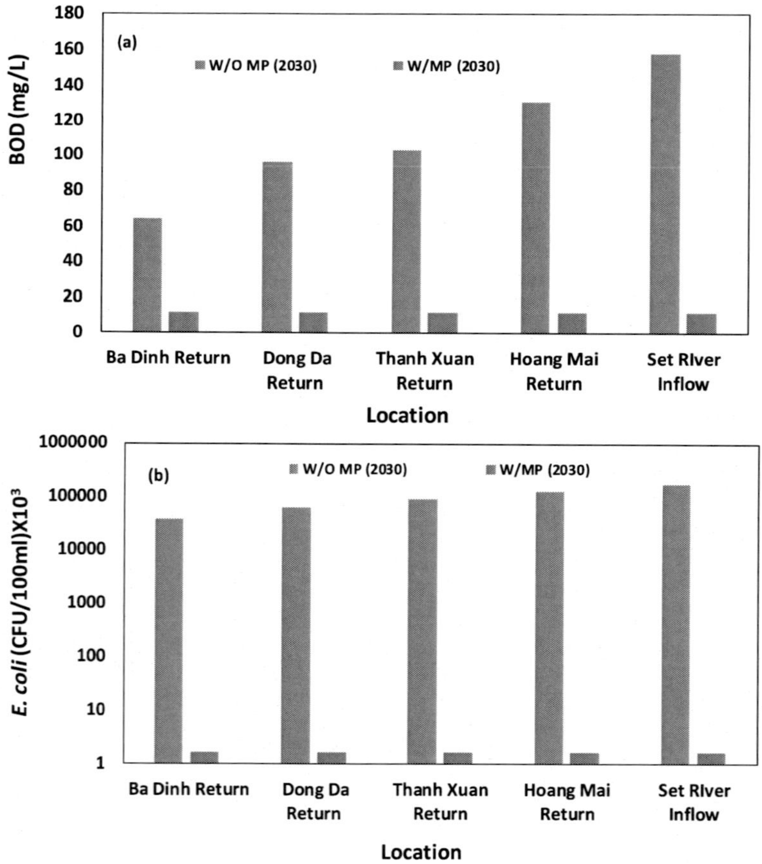

Figure 6. Simulated water quality result for (a) BOD and (b) E.*coli* in scenario with measures where all wastewater will be diverted from To-Lich River and treated wastewater from Ho Tay WWTP will be transferred to To Lich.

CONCLUSION AND RECOMMENDATIONS

This work strives to present a) current water quality status of all four rivers (year 2015), and (b) predicted the future water quality for To-Lich River (year 2030) using scenario analysis in Hanoi City, Vietnam. Results from this work suggests that water quality at present is worse in case of downstream sampling locations of most of the rivers. Looking into water quality guidelines, many of the location especially along the downstream don't comply with Class B for agriculture and aquaculture (BOD < 5 mg/L) (Ministry of Natural Resource and Environment, 2015). Also, with hydrological simulation, water quality will be deteriorating further in 2030 when compared to current situation. Due to climate change and population changes, the water quality (expressed as BOD) will be further deteriorating by 53.1% on average in 2030 compared to the current situation.

According to their master plan, all domestic wastewater currently flowing into the rivers except for the Kim Nguu River will be transferred to new wastewater treatment plants. Henceforth, this study suggest that the rivers should be supplemented with treated waswater from surrounding wastewater treatment plants or water from other sources to keep the river flow and to improve the water quality as shown in the mitigation plan scenario (85.8% reduction compared to the business-as-usual scenario).

If this masterplan succeeds by 2030, there will be no treated wastewater released to To Lich, Kim Nguu, Lu and Set rivers. These rivers will get dry during dry seasons. However based on the mitigation scenario, where whole treated waswater from Ho Tay WWTP (22.8 MLD) will be trasferred to To Lich river, water quality will be much better throughout the stream. In order to revive all four rivers with class B water quality, the local government has to think for diverting some of the treated wastewater into these rivers to maintain their flows especially during the dry season. However, the flow rate is still a matter of research and it can be achieved after regular monitoring once implemented. The national integrated sewerage and septage management program shall be implemented on priority basis, considering various factors like population density and growth, global changes for both short and long-term measures.

Some of the other potential reasons behind, the poor status of water quality are: a) At the current stage, despite the considerable capacity of existing WWTPs, the wastewater coming to these plants is not sufficient because of poor sewerage collection rate or poor connection between each household and main sewerage lines. b) Lack of proper coordination between different actors/stakeholders involved in water management to implement the master plan (water infrastructure) in a timely manner.

Possible solutions to overcome these barriers can be: a) To create some political space where different stakeholders other than government agencies also have direct involvement to influence the governing processes and government decisions; b) Provision of some financial incentives to encourage local people to connect their households to main sewerage line; c) People awareness about the benefits of better water environment in terms of good health and good business opportunities by tourism; d) As creating different hard measure (wastewater treatment) is a financial burden for many developing countries. There should be more push to the local government to implement de-centralized WWTPs along with centralized WWTPs; e) Local people or local government should be encouraged to maintain the existing septic tanks at regular basis. On behalf of such practice, there should be some monetary incentives like in terms of tax exemption should be given to those people; f) In order to complete the exiting master plan by the given timeline, regular monitoring for the progress of implementation of master plan is highly recommended.

REFERENCES

Angelo, Dell' J., Rulli M. C. and D'Ordorico, P. 2018. The global water grabbing syndrome. *Ecological Economics* 143: 276–285.

Blanco-Gutiérrez, I., Varela-Ortega, C. and Purkey, D. R. 2013. Integrated assessment of policy interventions for promoting sustainable irrigation in semi-arid environments: a hydro-economic modeling approach. *Journal of Environment Management* 128: 144–160.

Dao, C. A.; Con, P. M.; Khai, N. M. 2010. Characteristic of urban wastewater in Hanoi City – nutritive value and potential risk in using for agriculture. *VNU J. Sci. Earth Sci.*, 26, 42–47.

Downing, T. E. 2012. Views of the frontiers in climate change adaptation economics. *WIREs Climate Change* 3: 161–170.

FAO AQUASTAT website. 2016. Food and Agriculture Organization of the United Nations (FAO). Website accessed on [2016/05/15].

Ferguson, L., Chan, S., Santelmann, M. V. and Tilt, B. 2018. Transdisciplinary research in water sustainability: What's in it for an engaged researcher-stakeholder community? *Water Alternatives* 11(1): 1-18.

Frija, A., Dhehibi, B., Chebil, A. and Villholth, K. G. 2015. Performance evaluation of groundwater management instruments: The case of irrigation sector in Tunisia. *Groundwater for Sustainable Development* 1, doi:10.1016/j.gsd.2015.12.001.

Gareth, P. et al., 2014. Attitudes to water in South Asia. Chatham House Report. Chatham House, The Royal Institute of International Affairs, UK. Vinset Advertising, New Delhi, India, pp. 114.

General Statistics Office and United Nations Population Fund. 2016. Vietnam Population Projection 2014 – 2049. Vietnam News Agency Publishing House, Ha Noi.

General Statistics Office, Socialist Republic of Viet Nam. 2017. Statistical Yearbook of Viet Nam 2016; Statistical Publishing House: Hanoi, Vietnam.

Goyal, M. K. and Ojha, C. S. P. 2011. Evaluation of linear regression methods as downscaling tools in temperature projections over the Pichola Lake Basin in India. *Hydrological Processes* 25: 1453-1465.

Heinemann, A. B., Hoogenboom, G. and Faria, R. T. 2002. Determination of spatial water requirements at county and regional levels using crop models and GIS: an example for the State of Parana, Brazil. *Agricultural Water Management* 52: 177–196.

Hoekstra, A. Y., Buurman, J., and van Ginkel, K. C. 2018. Urban water security: A review. *Environmental Research Letters* 13(5), 053002.

Huizinga, J., Moel, H. de. and Szewczyk, W. 2017. Global flood depth-damage functions. Methodology and the database with guidelines. EUR 28552 EN. doi: 10.2760/16510.

Ingol-Blanco, E. and McKinney, D. 2013. Development of a Hydrological Model for the Rio Conchos Basin. *Journal of Hydrological Engineering* 18(3): 340–351.

Khan, A. A., Gaur, R. Z., Diamantis, V., Lew, B., Mehrotra, I. and Kazmi, A. A. 2013. Continuous fill intermittent decant type sequencing batch reactor application to upgrade the UASB treated sewage. *Bioprocess Biosystem Engineering* 36: 627–634.

Kumar, P. 2018. Simulation of Gomti River (Lucknow City, India) future water quality under different mitigation strategies. *Heliyon* 4(12), e01074.

Kumar, P., Masago, Y., Mishra, B. K. and Fukushi, K. 2018. Evaluating future stress due to combined effect of climate change and rapid urbanization for Pasig-Marikina River, Manila. *Groundwater for Sustainable Development* 6: 227–234.

MONRE (Ministry of Natural Resources and Environment) report. 2015. Water resource status and planning for its future management. pp72.

Mukate, S., Panaskar, D., Wagh, V., Muley, A., Jangam C. and Pawar, R. 2017. Impact of anthropogenic inputs on water quality in Chincholi Industrial area of Solapur, Maharashtra, India. *Groundwater for Sustainable Development* http://dx.doi.org/10.1016/j.gsd.2017.11.001.

Pink, R. M. 2016. Introduction. In Water Rights in Southeast Asia and India. Palgrave Macmillan, US, 1–14.

Quan, P. Van; Furumai, H.; Kurisu, F.; Kasuga, I.; Ha, C. T.; Chieu, L. 2010. Van Water Pollution Characterization by Pathogenic Indicators in Water Runoff in the Downtown of Hanoi, Vietnam. *J. Water Environ. Technol.* 8, 259–268, doi:10.2965/jwet.2010.259.

Slaughter, A. R., Mantel, S. K. and Hughes, D. A. 2014. Investigating possible climate change and development effects on water quality within an arid catchment in South Africa: a comparison of two models, In: Ames, D. P., Quinn, N. W. T., Rizzoli, A. E. (Eds.), *Proceedings of the 7th International Congress on Environmental Modelling and Software*,

June 15-19, 2014, San Diego, California, USA, ISBN: 978-88-9035-744-2.

Tundisi, J. G. 2008. Water resources in the future: problems and solutions. *estudos avançados* 22(63): 7-16.

UN Water. 2017. Wastewater: The untapped Resources. Facts and figure. The United Nations World Water Report 2017. Italy, pp 83.

United Nations, Department of Economic and Social Affairs, Population Division (UN DESA). 2015. World Urbanization Prospects: The 2014 Revision. *(ST/ESA/SER.A/366),* pp 517.

United Nations World Water Assessment Programme (UN WWAP). 2017. The United Nations World Water Development Report 2017. Wastewater: The Untapped Resource. Paris, UNESCO.

VUWSW. 2015. Vietnam urban water supply and wastewater project. Masterplan for wastewater management for year 2030. Pp 54.

In: Sustainable Water
ISBN: 978-1-53617-335-2
Editors: V. Gnaneswar Gude et al.

Chapter 5

UNCERTAINTIES IN WATER SUPPLIES DUE TO CHANGING CLIMATE AND EXTREME WEATHER EVENTS

Arumugam Sathasivan**, Shashika M. Korotta-Gamage and Ataur Rahman

School of Computing, Engineering and Mathematics,
Western Sydney University, Penrith, NSW, Australia

ABSTRACT

Water is an essential resource for ecosystems, daily needs of humans and economic development. Rightly water is closely connected with food provision and energy production (Water-Food-Energy nexus) and hence uncertainties in water supply cannot be discussed in isolation. Currently several regions and countries are heavily impacted by the water challenges but the scope and complexity widely vary. At the current level of demand from economic activity and population, about 53% of the population are living in the places where water is scarce or or close to reaching water scarcity. Further, increasing population and progressive increase in the

* Corresponding Author's Email: S.Sathasivan@westernsydney.edu.au.

economic activities increase the water demand hence, undermines water scarcity. Even at current level of population and economic activity, managing the risks associated with extreme events are difficult. With the global warming, frequency of extreme events is on rise; as a result, severe damage to the basic livelihood of wider spectrum of people across the globe could be felt. While the climate models accurately predict the atmospheric temperature, it is still difficult to predict rainfall and hence this is the most important factor for uncertainty. These uncertainties can then lead to uncertainty in water quantity, water quality, ecological functions, agriculture (food production), energy production and drinking water availability. This chapter explores the above aspects systematically to provide a comprehensive review of the available literature. In addition, an effort is made to point possible adaptation measures to mitigate possible impacts.

Keywords: climate change, water quality, water availability, drinking water supply, agriculture, energy

INTRODUCTION

The scope and complexity of water challenges cut across regions and countries. The water scarcity – water demand is either higher or reaching the amount of water available – threatens socioeconomic development and national security for many countries around the world. The International Water Management Institute (IWMI) estimates that about 20% of the world population are experiencing water scarcity and about 8% live in areas that will soon approach water scarcity. About another 25% live in areas where there is an economic water scarcity – water is present, but these areas lack financial/human capital to access water (Bates et al. 2008). This situation cannot be attributed solely to natural phenomenon, humans have to take equal blame for the ways they use the water.

Human actions have negatively affected both quantity and quality of water. Poor management of water resources has deteriorated water quality to the extent that it greatly threatens human and ecosystem health, increases water treatment costs, and reduces the availability of safe water for drinking and recreational uses (Gleick and Palaniappan 2010). Globalization in trade has facilitated "virtual" water movement across the globe. For example when rice – a product of high water demanding crop - is exported from one country to another, the water requirements in the importing country reduces but at an expense of increased water demand and deteriorated water quality in the exporting country. Not recognizing the Food-Water-Energy nexus amongst food, water and energy has also contributed to issues with deterioration of water quality and quantity. Agriculture consumes around 70% of water resources but in some places of US, energy production has contributed to 50% of freshwater withdrawals for thermoelectric power plants and thus the increased polluted water disposal. Economic development of underdeveloped countries, urbanization and population growth will increase the appropriation of water thus increasing the water stress.

The uneven distribution of water resources and deterioration of water quality represent bigger problems compared to the concerns associated with reduced availability of water in the world. In addition to well-recognized factors including population growth, urbanization, poverty, and economic development, climate change adds a new complexity to existing issues associated with scarcity of water supplies. The predicted increases in frequency and intensity of extreme weather events, including heat waves, drought, hurricanes, wildfires, and floods (Figure 1) could place enormous strain on water supplies, agriculture, and the built environment. The climate change has already been reported to play a pivotal role in the re-emergence of pathogens and spread of diseases to new regions (Funari, Manganelli, and Sinisi 2012). These trends pose urgent threats in developing and developed countries alike.

Recognizing the multifaceted water challenges facing the humanity, several international and national agencies have incorporated "access to clean water and sanitation" as one of their top strategic priorities. The UN has incorporated the water challenge – ensuring availability and sustainable management of water and sanitation for all - as the sixth Sustainable Development Goal.

Further, other goals such as zero hunger (Goal 2), good health and wellbeing (Goal 3), affordable and clean energy (Goal 7) are invariably related to the provision of water. National Academy of Engineering (NAE) of USA has identified "provision of clean water" as one of the fourteen "Grand Challenges of Engineering". In providing the vision for the future of Environmental Engineering education, NAE calls for "holistic consideration of the consequences of our actions in developing the solutions that will be needed in the coming decades. For example, the development of biofuel (water intensive crop production is needed to generate raw material) as an alternate energy source is expected to aggravate water and food security issues.

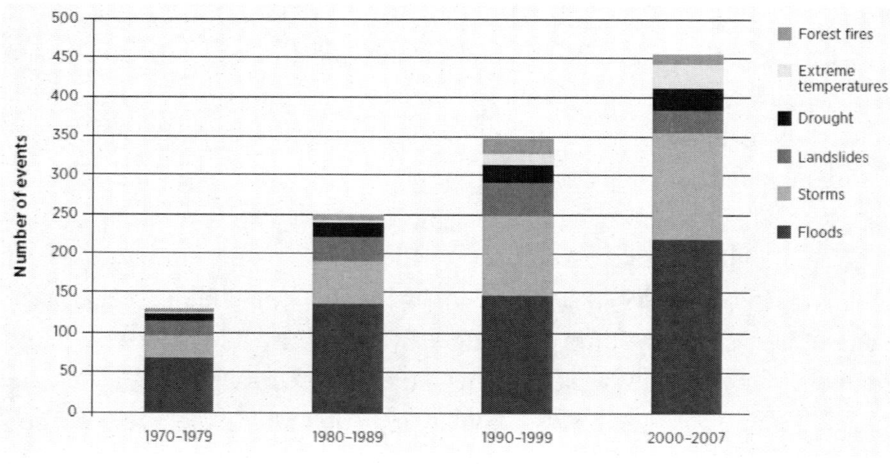

Figure 1. Frequency of hydrometeorological events from 1970–2007 (United Nations 2012).

Based on the above background, the role of this chapter is not to address every aspect of water challenge but primarily explore how climate change and extreme events negatively impact water supplies. This is done by addressing how water availability is affected quantitatively and qualitatively by the climate change and how water supply to different water intensive sectors are affected by the uncertainties arising from climate change and extreme weather events such as prolonged drought, flood, or cyclone (typhoon).

IMPACT ON WATER AVAILABILITY

Water availability in a given region is likely to be affected by global warming in a number of ways. The warming of the atmosphere and ocean, which happens at a non-uniform rate over space, has been impacting the spatial distribution of rainfall. Many wetter regions are getting drier and some drier regions are becoming wetter. For example, Hajani and Rahman (2018) have reported that the rainfall intensities of longer duration events in New South Wales, Australia have a decreasing trend during the last 30 years. Ishak et al. (2013) found that peak runoff has a decreasing trend in south-east Australia, while the Northern Australia is experiencing an increasing trend. Al Mamoon and Rahman (2017) found that Qatar has been experiencing a decreasing trend in its annual rainfall. Hossain and Rahman (2019) found that Brisbane river catchment is showing a decreasing trend in annual maximum flow rate.

Inter-event durations of bigger rainfall events are increasing, i.e., the duration of drought is increasing in many regions. The higher variability in rainfall driven by climate change implies that the reservoir size will need to be increased in order to enhance the long-term reliability of water supply. Haque et al. (2015) found that catchment water yield has a decreasing trend in Blue Mountains catchment in New South Wales, Australia. Similar finding was reported by Chiew et al. (2009).

The efficiency of rainwater harvesting will also be affected by climate change as found by Haque, Rahman, and Samali (2016). Climate change also affects the catchment wetness index, which in turn affects the runoff generation (Sharma, Wasko, and Lettenmaier 2018). Due to increased temperature, the evaporation rate is on rise, which reduces the volume of available water from water supply reservoirs, lakes and rainwater tanks. Evaporation rate has a significant impact on water availability due to climate change. Climate change (warming) will increase the evaporation at a greater rate in arid regions as compared to the humid regions. Water use in construction will increase due to higher evaporation loss as noted by Rahman et al. (2019).

Other impacts of climate change are increased heat wave and urban heat island effect, both of which make air drier and increase water demand for drinking and recreational uses. The numbers of days exceeding a certain threshold temperature (both maximum and minimum) have been increasing, which increase the discomfort level of people, which eventually increase water demand, as people tend to have a longer shower during hot weather (Ouarda and Charron 2018). The frequency and severity of bush fire have been increasing due to climate change, which increase the fire-fighting water demand and water quality.

Due to increasing temperature, the snow melting has been increasing, which will increase the streamflow in the melting season. This increased streamflow can enhance the reservoir water inflow in some regions if an appropriate reservoir exists in the area or a new one is built.

The sea level rise will affect many island nations in a number of ways, such as increased surface and groundwater salinity. This will make water unsuitable for drinking and irrigation. People, in particular, female household members will have to travel a long distance to fetch drinking water to meet the family needs. As an example, the salinity level of coastal regions of Bangladesh has been increasing rapidly and it has been reported that people need to travel over ten kilometres to fetch drinking water at many locations in southwestern Bangladesh.

The drainage of agricultural land will be affected due to sea level rise and water logging problem will increase at many coastal locations, and hence water demand during dry season will increase to make up the lost crop production in rainy seasons.

Floods are on the rise as the frequency and severity of extreme rainfall has been increasing (Laz et al. 2014) at many locations and hence more extreme flood events are increasing in numbers. These floods increase erosion, damage crops and increase the water availability over a short period of time, giving opportunity to store this excessive water. However, to store the flood water, more reservoirs need to be built, in particular at the upper part of the catchments; however, in plain land, near the mouth of the sea, this is not feasible as the reservoir will need larger surface area due to smaller height difference.

Groundwater is a primary source of water for both drinking and agricultural purposes in many rural communities. Unsustainable groundwater extraction has already increased the risk of salt-water intrusion in many parts of India and China. Sea level rise will aggravate the salinity intrusion for many coastal aquifers, making the water unsuitable for drinking and irrigation. This will put more pressure on alternative water sources such as surface water bodies or harvested rainwater.

IMPACT ON WATER QUALITY

Water quality is inseparably linked with water quantity, as both are key determinants of water supply for basic human and environmental needs. As identified by the United Nations in 2012, poor water quality not only impacts on health and well-being of humans but also the social and economic development and sustainability of ecosystems and biodiversity (United Nations 2012). Moreover, water quality issues have brought forefront of international action by addressing in 2030 Agenda and by setting Sustainable Development Goal (SDG) 6 to respond the challenges posed by water quality issues (United Nations 2018; UN Water 2018).

Warming of the earth's climate in recent decades as a result of rising concentration of greenhouse gases are now evident from the observations of increase in global average surface temperature by about 1.8 °F (1.0°C) over past 115 years at an increasing rate, accompanied by widespread melting of snow and ice, rising sea levels, resulting extreme weather events. Earth's temperature is also expected to increase by 4.7°F to 8.6°F (2.6°C to 4.8°C) by 2010 due to continuous emissions of greenhouse gases (National Academies of Sciences 2019). These changes in the atmospheric temperature and radiation balance are intimately linked with hydrological cycle and they can alter the phenomena, which invariably affect the water quality such as precipitation patterns and intensity, evaporation, soil moisture and runoff. Indeed, major international and national assessments of climate change have identified 'water and its availability and quality will be the main pressures on, and issues for, societies and the environment under climate change' (Bates et al. 2008; U.S. Global Change Research Program (USGCRP) 2018). Therefore, the relationship between climate change or extreme weather events and water quality of freshwater resources is the primary concern and interest. Thus, the main water quality parameters, which are highly affected by the rising temperature and extreme weather events such as drought, heavy rainfall, storms and hurricanes, are discussed herein.

Dissolved Organic Carbon

Natural organic matter (NOM), a measure of dissolved organic carbon (DOC) is a major reactant and product of biochemical processes. DOC that abundantly occurs in all water sources serves as carbon and energy source for associated biota. The origin of NOM can be autochthonous due to the activity of algae, cyanobacteria in water or allochthonous; terrestrial origin dominated by humic substances (Pivokonsky et al. 2015). The presence of DOC in portable water is aesthetically undesirable; cause colour, tastes and odour problems and contribute to disinfection by-products (DBP) formation and associated with health risks (Kastl, Sathasivan, and Fisher 2016).

Evaluation of DOC concentrations for 198 sites across the UK including streams, lakes, and water supply reservoirs over eight years had observed 0.17 mg-C/L average annual increase in DOC concentration (Worrall et al. 2004). Similarly, due to the range of potential drivers including temperature, rainfall, acid deposition, nitrogen and carbon dioxide enrichment, increase in DOC concentration on an average 6.1% per annum or 91% over 15 years (from 2004) was also observed in 22 UK upland waters (Evans, Monteith, and Cooper 2005).

Most of the recent research have reported the transport of terrestrial DOC from soil to streams due to extreme weather events such as high rainfall (flooding), tropical storms, hurricanes and snowmelt events, and increase the concentration of DOC in downstream aquatic systems including coastal rivers and estuaries (Bianchi et al. 2013; Hounshell et al. 2019). According to the pulse-shunt concepts (PSC) proposed by Raymond, Saiers, and Sobczak (2016), extreme weather events trigger a ‘pulse’ that releases terrestrial DOC and exported from land to adjacent streams and rivers and then the same event accompanied by a rapid increase in headwater stream velocity and ‘shunted’ a large portion of the pulsed DOC further to downstream. Extreme flood events like Hurrican Matthew (in 2016) exported over 99% of terrestrial DOC load from the riverine end-member to the downstream coastal system (Hounshell et al. 2019). A research study focused on 30 small eastern United States forested watersheds reported 86% of DOC was exported to streams during hydrological events such as storms and snowmelt (Raymond and Saiers 2010). Further, in a larger context of climate change, frequency and intensity of extreme weather events are predicted to increase and this projected to change the frequency and intensity of extreme precipitation events (Janssen et al. 2016).

Concurrently, reduced rainfall levels and drought events impacts the mobilization and accumulation of material due to the reduced flow rates or flow velocities, and alter the chemistry and treatability of NOM (Wright et al. 2014). Moreover, temperature rises reported to increase the DOC levels in water sources primarily attributed to enhanced enzymatic activity in microorganisms and enhanced biodegradation of soil organic matters (Pagano, Bida, and Kenny 2014; Lipczynska-Kochany 2018).

Nutrients

Climate changes accompanied by extreme weather events are predicted to affect the hydrological cycle which could modify the nutrient criteria, nutrient transformation and transport process in terrestrial and aquatic environments. These changes in the nutrient patterns eventually, has altered the relationship between nutrients and primary productivity or the ecosystem functions. A warmer climate or rising temperatures are expected to increase the rate of mineralisation of soil nutrients (nitrogen, phosphorous and carbon), rate of nutrient release i.e., nitrogen and phosphorus into overlying water and deoxygenation at the lake sediment surface (increase the release of phosphorus) (Kosten et al. 2012; Me et al. 2018). Further, the higher temperatures and increased emissions of long-wave radiation increase lake water temperature at the surface and stimulate more persistent stratification which in turn modify the nutrient dynamics; mainly, higher ammonium and phosphate levels (Arnell et al. 2015). During droughts under low-flow river conditions increased concentrations of ammonium, nitrite and orthophosphate was observed due to the reduced dilution effects of effluents of wastewater treatment pants (van Vliet and Zwolsman 2008). A modelling study of three New Zealand lakes (Trolle et al. 2011) showed that the effect of climate change by 2100 will equate to increasing external nutrient loads by 25-50%.

An increase in the total phosphorus loads of 3.3% to 16.5% in Denmark streams in the next 100 years was predicted by Jeppesen et al. (2009) in response to the increased precipitation in the winter due to higher winter rainfall and to a decline in warm temperate.

In addition, extreme weather events such as heavy rainfall after the drought period and large storms flushed greater nutrient inputs into water bodies causing higher nitrate and phosphorus concentrations. Total phosphorus concentration in Tuross estuary from the Tuross River catchment of southeast Australia raised to a maximum of 0.25 mg/L during July 2005 largest storm event (Drewry, Newham, and Croke 2009).

As a result of increasing nutrient input via increasing water temperature hydrologic changes, more intense and persistent droughts heavy rainfall increase the growth of cyanobacteria relative to other algae and promote the formation of more harmful cyanobacterial blooms (Kosten et al. 2012; Huo et al. 2019) and deteriorate the water quality. Further, at higher water temperatures, DOM originating from algal sources count for about 20-50% of total DBP formation potential (Delpla et al. 2009).

Human on the other hand can reverse negative impacts, showing these could be adopted as a solution. In river Rhine, Switzerland ammonium concentrations in 1976 drought were an oder of magnitude higher than in 2003 drought, thus, with the progressive improvement of wastewater treatment observed nutrient concentrations were reduced from 1976 to 2003 drought (Zwolsman and Van Bokhoven 2007)..

Micropollutants

Micropollutants occur in the aquatic environment and are a matter of growing concern worldwide due to its harmful effect on humans and biota. The micropollutants comprise of a wide range of natural and synthetic inorganic and organic compounds, which include pharmaceuticals (i.e., antibiotics, steroid hormones, antidepressants, lipid regulators, nonsteroidal anti-inflammatory drugs) and personal care products (PPCP) (i.e., sunscreen, soaps, moisturizers, lipsticks, fragrances, insect repellent, and shampoo), detergents, industrial chemicals, pesticides, and many other contaminants. Thus, micropollutants are anthropogenic chemicals that occur in the aquatic environment at trace levels (pg/L, ng/L) due to human activities. The climate changes and climate extremes such as flood events play a key role in increasing the contamination of freshwater via redistribution of chemical micropollutants. Further, flood events assumed to enhance the release of micropollutants stored in ice, soil or sediments (Whitehead et al. 2009). Higher rainfall increases the runoff which is the lateral movement of water either over the surface or through the top layers of the soil to a watercourse. In extreme rainfall events, up to 5% loss of the applied dose has been measured in the soil and these small amounts can lead to higher pesticide concentration in drinking water (Bloomfield et al. 2006).

During drought events discharges from point sources are less diluted thus, will increase the concentration of compounds. Higher concentration of certain heavy metals and metalloids i.e., barium, selenium and nickel were observed in river Meuse western Europe during the drought of 2003 (van Vliet and Zwolsman 2008). Similarly, Osorio et al. (2012) and Masiá et al. (2013) observed a higher concentration of pesticide and pharmaceutical compounds in surface water during low river flows compared to high flows. Moreover, mountain topography (snow fields) support the deposition of different compounds at different altitudes and the formation of hotspots of micropollutants (Schmeller et al. 2018). Therefore, many rock glaciers in mountain areas are sensitive to climate warming and thus, mountain freshwater becomes increasingly contaminated by micropollutants as Thies et al. (2007) observed in alpine lake water (European Alps). In drinking water derived from meltwater of glacier, a very high concentration of nickel (48-67 μg/L) was observed at high alpine refuge Becherhaus (Thies et al. 2007).

However, the fate of micropollutant in relation to long-term climate change or extreme weather events remains poorly understood in comparison to macropollutants such as nutrients, natural organic matter, salts and acids (Schmeller et al. 2018).

Pathogens

Climate changes accompanied by extreme weather events are among the main factors that impact waterborne pathogen concentration in the worldwide surface water and thus, increase the waterborne diseases. More frequent and intense heavy rainfall or floods can cause over flooding of sewage treatment plants, sewer overflows, runoff of animal manure, re-mobilisation and redistribution of contaminated sediments and cause a high concentration of pathogens and nutrients in natural water sources. Waterborne disease outbreaks associated with excessive rainfall was reported worldwide and few examples are drinking water contamination with *Cryptosporidium* cists in Milwaukee (USA) in 1993, groundwater contamination with *Acanthamoeba keratitis* in Iowa (USA) in 1994, *Escherichia coli* and *Campylobacter jejuni* waterborne outbreak in Canada and diffusion of cholera in many developing countries (Funari, Manganelli, and Sinisi 2012).

The impact of increased temperature on pathogens is not expected to be homogeneous. Higher water temperatures could probably favor less temperature sensitive species, on the other hand, inactivate temperature sensitive viruses, bacteria and parasites. Increasing temperature expected to increase the prevalence of *V. cholerae* and cholera disease in Asia and South America (Hunter 2003). In addition, increased precipitation events during the droughts increased the concentration of waterborne pathogens in the water (Hofstra 2011).

Uncertainties in Water Quality

The hydrological cycle is driven by a complex, interrelated, dynamic natural processes. The quality of water resources is one of constant change as a result of variability of the earth's climate such as increased flood potential, increased evaporation losses, changes in seasonality or timing, changes in flow from glaceirs, decresing snow and permafrost, groundwater depletion and changes in soil moiture (United Nations 2012). These source of change and the interaction between them create high level of uncertainty associated with the availability and quality of water resources.

Concurrently, there is high uncertainty associated with projected or modelled changes in the hydrological system arising from future greenhouse gas emission, climate changes or variability in weather events. Moreover, by the late 21st century, under the projected scenario, there is high uncertainty with climate model precipitation projections which is reflected in a wide range of climate model estimates of future precipitation (Bates et al. 2008). These uncertainties in global climate model projections lead underestimation of future extreme weather events. Similarly, other changes in water quality such as pollutant load, nutrients, pathogens and sediments are associated with high uncertainty due to a combination of uncertain hydrologic processes (U.S. Global Change Research Program (USGCRP) 2018). Eventually, these uncertainties in water quality affect on water treatment and supply.

IMPACT ON DRINKING WATER PRODUCTION AND SUPPLY

Providing water in sufficient quality and quantity is essential for all aspects of life and sustainable development. However, the world is facing increasing problems in providing safe drinking water services due to population growth, urbanisation, industrial production and pollution and climate changes. Estimates suggest that if the natural environment continues to be degraded and unsustainable pressures put on global water resources, 52% of the world's population will be put at a risk by 2015 (UN Water 2018). The additional burden provides by climate changes on water utilities in providing safe drinking water is represented in this context.

Impact on Coagulation

Coagulation using aluminium or iron salts followed by flocculation and filtration is the most commonly used method in drinking water treatment to remove DOC. The efficient removal of DOC by coagulation relies on establishing optimum conditions such as coagulant dose, pH and mixing time (Kastl et al. 2004) and these conditions depend on the raw water quality including DOC concentration and characteristics (proportion of hydrophobic/hydrophilic fraction), pH, alkalinity, turbidity and temperature.

The climatic changes coupled with increasing production of DOC increased the coagulant dose requirement which is associated with more sludge production. Recent studies have shown that DOC produced under the drought conditions are more hydrophilic, hence, harder to remove by the coagulation process (Tang et al. 2013). On rewetting, the ratio of hydrophilic/hydrophobic fraction of DOC decreases (Fenner et al. 2001) and this hydrophobic fraction of DOC is easily removed by the coagulation. However, drought events caused an adverse effect for water treatment and long-term issues as the hydrophobic fraction of DOC can remain elevated for some years (Watts et al. 2001) and increased concentrations of both hydrophobic and hydrophilic fraction of DOC act as more health concerning DBP precursors.

Similarly, the increase in DOC during the heavy rainfall and storm events cause issues for water treatment relating to process optimisation and rapid response to changing DOC quality and quantity. In addition, increased in turbidity or dissolved and particulate solid concentrations, debris and excess nutrients during the flood and storms cause more severe problems in the drinking water treatment process (Ritson et al. 2014).

Impact on Disinfection

Elevated DOC concentration affect the water treatment process in many ways; as DOC can increase the precursor for the harmful disinfection by-products (DBP) produced during the disinfection, promote the microbial growth and biofouling. Increased concentration of DOC remaining after coagulation serves as the precursor for the DBP formation. Whereas, the higher DOC levels and microbial growth will increase the disinfectant dose because of higher demand and need for adequate residual levels and this further increase the DBP formation. Moreover, Hua and Reckhow (2008) reported higher temperature during the disinfection process can increase the concentration of DBP.

Global warming induced sea level rise can cause saltwater intrusion into coastal aquifers, as reported in the Cyprus and Gaza Strip (Stellar 2010), Bangladesh, Australia, Brazil and Netherlands (Winid 2015). Generally, seawater contains around 65 mg/L of bromide which is much higher concentration than the freshwater (Magazinovic et al. 2004). As a result, a rapid increase in bromide concentration can be observed in water sources in coastal areas. Disinfection of water containing bromide can shift DBPs from chlorinated to more brominated species and this tends to increase with the bromide concentration. These brominated DBPs are more cytotoxic and genotoxic than their chlorinated analogues (Winid 2015).

Even though the chlorination is an essential process in drinking water treatment to inactivate pathogenic microorganisms, exposure to more health concerning DBPs formation has to be considered under high DOC concentrations.

Risk in Desalination Techniques

As discussed in Section 2, climate drivers such as precipitation, temperature and evaporation demand significantly impact on freshwater availability. In the context of freshwater deficiencies, desalination technologies are now being developed worldwide. In year 2015, about

18000 desalination plants have been established wroldwide mainly, in Middle East and North Africa and produced nearly 23 billion gallons of fresh water per day using reverse osmosis and distillation technologies (National Academies of Sciences 2019). However, desalination is too expensive and energy intensive to serve as a widespread solution for water scarcity.

Desalinated water is sometimes the only alternative for the water scarce countries close to coastal areas. However, disinfection of salinated water give rise to formation of a wide range of potentially carcinogenic organobromine compounds (Winid 2015). Therefore, rigorous risk control measures have to be implemented for drinking water treatment methods to provide water that is safe for public health.

Health Impacts

Observed trends in deteriorating drinking water quality under the changing climate changes and extreme weather events impact on the drinking water treatment and supply by increasing the health risks. The World Health Organization (WHO) estimated that in 2012, 12.6 million deaths worldwide, represent 23% of all deaths were attributable to the environment (World Health Organization (WHO) 2016). Mainly, increased DOC levels in surface water and elevated temperatures significantly increase the formation of more health hazardous DBPs. Concurrently, the presence of bromide and iodide during disinfection lead to shift the proportion of DBP formed from chlorinated species to more brominated and iodinated compounds which are more toxic than the chlorinated DBPs.

In Bangladesh, approximately 35 million of coastal populations are highly vulnerable because of salinization of fresh water as a result of climate change and sea level rise. Expose to saline-contaminated water through drinking, increases the potential of health impacts including skin diseases, miscarriage, hypertension, diarrhoea, and acute respiratory infections (Rakib et al. 2019).

In addition, the change in the discharge water quality due to the varability of climate changes such as flood and storm events, affect the incidence of the waterborne pathogens such as feacal indicator bacteria; *E. coli* and at higher temperatures toxic producing algal blooms pose greater problems in water treatment (Mohammed, Tveten, and Seidu 2019). Further, the presence of micropollutants in the surface water increase the health risk and have a huge impact on drinking water treatment. Even if they are partially removed, residual quantities remain in drinking water have been found to be highly toxic (Hernando et al. 2006).

Economic Aspects

Safe access to drinking water depends on the level of water supply infrastructure. However, climate changes lead to additional costs for the water treatment and supply sector due to changing water levels and quality such as increasing DOC, pollutant and microbial contaminants.

When freshwater supply has to be replaced by desalination water due to climate change, the average cost for desalination is around US$ 1.00/m^3 for seawater and US$ 0.60/m^3 for brackish water (Bates et al. 2008).

However, developing and implementing new drinking water treatment approaches pose a special challenge to developing countries and other rural communities with limited financial and technical resources and thus, become more vulnerable for climatic impacts.

Impact on Water Supply for Agriculture and Energy

Agricultural Water Supply

Key variables that govern plant growth are nutrients, temperature, availability of water to satisfy evaporative demand and other demands such as maintenance of soil moisture at optimum level. Water availability from water resources (rainfall, natural surface water bodies and ground water) is a critical factor for agriculture and is expected to change as a result of climate change. To understand the impact, prediction models are crucial. However, it is well recognized that the long term risk to agricultural production cannot be predicted with certainty, because of uncertainty in water availability. Global circulation models predict temperature and pressure accurately but the water vapor - the important determinant of rainfall – cannot be as confidently as temperature and pressure. This uncertainty may not affect the other sectors as they consume less amount of water as opposed to agriculture which consumes as high as 70% of total water withdrawals in the world (WWAP (United Nations World Water Assessment Programme) 2014). In addition, the unavailability of models with sufficient accuracy at regional level makes it even harder to predict the likely scenarios. This makes it difficult to assess the required level of investment on irrigation infrastructure accurately.

Irrigation (from canals and water ways) provides approximately 40 percent of the world's food, including most of its horticultural output, from an estimated 20 percent of agricultural land, or about 300 million ha worldwide (Turral, Burke, and Faurès 2011). Change in precipitation can seriously affect the canal/river flows depending on the location. For example, in Morocco, 5% fall in precipitation will reduce the run off by 25% (Turral, Burke, and Faurès 2011).

By 2050, the population will reach nine billion with majority of the cities urbanized (WWAP (United Nations World Water Assessment Programme) 2014). Urbanized population usually consumes high water demanding food such as meat and dairy products. One kg of beef require around 15 tons of water as opposed to 5 tons of water for one kg of rice (Chapagain and Hoekstra 2008). Although some water efficiency (about 20%) in agriculture is possible, the increased demand from population and life style changes due to economic development can easily outweigh the efficiency.

There is sufficient water across the world, but it is the uneven distribution, different agricultural practices and uneven effect of climate change across the globe that make the affect much more pronounced. For example, the drought in coastal area and heavy reliance of ground water for agriculture and drinking/industrial needs can make groundwater depletion and hence salt water intrusion as currently experienced in many parts of Southeast Asia. The sea level rise will aggravate the salt water intrusion. On the other hand, increased rainfall and flood in wetter regions of tropics and subtropics can bring excess amount of water that cannot be practically captured safely within the short time and suitable infrastructure could not be built. The flood also causes soil erosion and hence nutrient depletion in the soil.

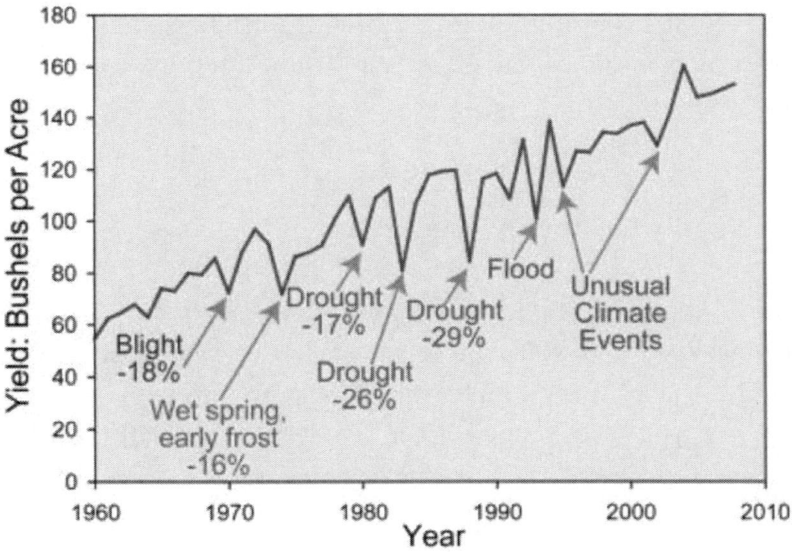

Figure 2. Impact of extreme weather events on yield reductions (United States Environmental Protection Agency (USEPA) 2017).

Increase in temperature up to an optimum level and that of carbon dioxide can increase the photosynthesis and thus the production, but beyond an optimum level the plant can start to die (United States Environmental Protection Agency (USEPA) 2017). Increased carbon dioxide level will likely to assist in increasing the production, but after the optimum temperature, the plant will die anyway negating the positive impact. The algae are not much different to the plant in terms of impact from temperature and carbon dioxide levels. In addition to this, nutrient supply from the cycling of nutrients in water bodies will increase the nutrient supply to increase the productivity of algae and hence the dissolved oxygen levels in the night can substantially decrease in the water bodies which heavily impair the fisheries and aquaculture. Increased water temperature may also alter the species or the life cycle of aquatic organisms (United States Environmental Protection Agency (USEPA) 2017).

While climate change impacts can induce the above changes the other stressors such as sudden events - drought, flood and cyclone will heavily reduce the productivity and hence the socioeconomic status of a country. The increase in population and heavy reliance on agriculture in developing countries for income will aggravate these negative impacts and uncertainties.

Water Supply for Energy Production

Global energy production is expected to increase by 60% by 2035 and approximately 90% of global power generation is water intensive. Next to agriculture, global water withdrawals for energy production is the major portion - around 15% of total water withdrawals in 2010 (International Energy Agency (IEA) 2012). Despite ongoing progress on the renewables, fossil fuels (including thermal power plants) still provide most of the energy demand but increased use of biofuels which are water intensive can substantially increase the water demand. Thermal power plants use the maximum amount of water. Combining with the increased use of biofuels (water and land intensive), the water withdrawals by the energy sector can increase to 35% (without efficiency in cooling systems etc. the increase will be even more).

High current and future water demand can substantially compete with other uses of water such as agriculture, environmental flow, drinking and other industrial activities. Uncertainties in global energy production mix along with climate change induced uncertainties can pose substantial risks to water and food supplies. Recently to cater for high oil prices, biofuels were encouraged which impacted directly on the food available for consumers making food prices to increase and stress on water resources. In a fragile environment with only a slight amount of water available for environmental flow, uncertainty generated by the rainfall will substantially increase the tension between different uses.

Hydropower is seen as the best buffer or insurance against the fluctuating power demand from climate change and extreme drought. Previous studies on the impact of climate change on hydropower show that results are very site-specific (Liu et al. 2016; Majone et al. 2016). Even neighboring catchments may face opposite impacts of climate change (Gaudard et al. 2014) and thus completely different risks.

Adaptation to Climate Change and Extreme Weather

Many regions around the world already experience significant impacts from climate change and these are projected to increase over the time. Therefore, the challenge is not only to reduce greenhouse gas emission but also of adapting to current and future climate change risks. As defined by U.S. Global Change Research Program (USGCRP) (2018), adaptation refers to action taken at the individual, local, regional and national levels to reduce risks from even today's changed climate conditions and prepare for impacts from additional changes projected for the future. Thus, adapting to uncertainties in water supply with the increasing climate change threats requires a mix of measures including implementing new technologies, infrastructure and physical installations as well as changes in water management policies.

Provide Safe Drinking Water

Providing safe drinking water is the most challenging aspect due to severe contamination (micropollutants, pathogens) and highly variable quality of freshwater sources (increase concentrations of DOC, nutrients) and lack of access to infrastructure due to physical location, financial capability and many climate influenced effects. Moreover, due to varying scenarios for different weather events such as storms, flood, drought, heat waves, it is impractical to adopt the same solution in every location for every change in the climate or weather events. The minimum standard required for drinking water for disaster response is no faecal coliform per 100 mL, free chlorine residual > 0.2 mg/L, turbidity <5 NTU and no negative impact on health due to short-term consumption.

Loo et al. (2012) proposed ten mutually independent criteria to evaluate the appropriateness of water technology for a given emergency. These criteria include costs, ease of deployment, ease of use, environmental impact, maintenance, performance, potential acceptance, energy requirement, supply chain requirement and throughput. Scoring system developed based on these ten criteria suggested that a judicious choice of the relevant water technology to be deployed for emergency relief is crucial.

Membrane technologies such as ceramic filter, ultrafiltration are commonly used in disaster relief due to their high microbial and turbidity removal efficiency. Another water technology that has good potential for disaster relief is combined coagulant and disinfectant powder; PuR® sachet, which is suitable for household level intervention (Loo et al. 2012). The PuR® is capable of producing microbiologically safe drinking water and has an added advantage of removing 99.8% arsenic. The use of PuR® has reported in diarrhea reduction of 90%, 24% and 19% in Liberia, Guatemala and Kenya, respectively (Loo et al. 2012). Further, adsorbent technologies such as biologically activated carbon (BAC) can be used to remove toxic substances including micropollutants organic and particulates (Korotta-Gamage and Sathasivan 2017).

Infrastructure for Better Storage and Distribution

Water use is increasing around the world with the population growth, economic development and lifestyle changes. Concurrently, water availability is likely to decrease with the influence of climate changes; global warming induce sea level rise-salt water intrusion and extreme weather events such as drought. At the same time, increased runoff is expected as a result of extreme weather events such as storms and flood events. Therefore, adaptation approaches should focus on both conditions of water scarcity and water excesses.

Most developing countries are still in the process of meeting the basic levels of water infrastructure development. However, there is a substantial requirement for increased water storage with the uncertainties in rainfall projections.

Construction of water storage facilities; dams, reservoirs or the exploitation of natural systems; enhancing groundwater recharge through rainwater harvesting, sustainably managing aquifers and nourishing wetlands allow water to be stored for use of during drought periods and control the damage caused during floods (United Nations 2012). Further, infrastructure for proper water distribution and canal system will be required in many regions to guard against severe flood events and changes in river flows and to replenish the depleted ground water resource from over exploitation. Creating infrastructure for water distribution will provide additional benefits for irrigation, surface and groundwater storage and prevent the flowing of excess water directly to the sea. However, the new structure should only be built only after optimization of existing infrastructure and proper analysis. Further, the canal irrigation systems which supply sufficient water will encourage the farmers to reduce the groundwater exploitation and thus slow down the ground water depletion.

More rigorous water balance and catchment yield models needs to be developed which are capable to make realistic assessment of the impacts of climate change on water availability at a given location. These models should be able to simulate the future drought frequency and severity in a more realistic fashion. Traditional linear and stationary water balance/yield models cannot achieve this realistic future prediction. The results of climate change based hydrologic modelling will assist to upgrade our existing reservoir systems to hold more water to meet the demand during the prolonged dry periods, which are more likely to be a regular event in near future.

More efficient water management methods are to be developed such as adoption of sub-surface irrigation systems to replace flood irrigation methods. The concept of protected cropping is to be encouraged which needs less water and increases agricultural production. Smart water use practice in the middle-eastern countries like Israel can help many water stressed countries to learn how to cope with little water.

More efficient water management methods are to be developed such as adoption of sub-surface irrigation systems to replace flood irrigation methods. The concept of protected cropping is to be encouraged which needs less water and increases agricultural production. Smart water use practice in the middle-eastern countries like Israel can help many water stressed countries to learn how to cope with little water.

To avoid the movement of virtual water (especially the water hungry crops such as rice, cotton) across the globe, water hungry plants such as rice should only be encouraged in areas where there is abundance of water.

In some cases, recycled wastewater and desalinated water could be used or water efficiency improved such as that practiced in Israel.

Water quality requirement for agriculture is not as serious as that of drinking water, but saline (from the sea or ground water) or acidic water (from acid rain or exposure to acid sulphate soils) or that contaminated with heavy metals such as arsenic compounds cannot be utilized. Continuous use of recycled wastewater containing high concentration of dissolved solids have shown to increase the salinity of irrigated soils.

Coherent Policy-Making

Substantial changes have been reported over recent decades around the world in many water-related variables and further variations are projected in future in many regions. For example, the per capita availability of freshwater in India is expected to drop from around 1820 m^3 currently to below 1000 m^3 by 2025 in response to the combined effect of climate change and population growth. Projected warming in the western mountains in North America by mid 21st century is expected to cause a large decrease in snowpack, earlier snowmelt, more winter rain events, increase peak winter flows and flooding (Bates et al. 2008). Therefore, coherent policy-making is essential to facilitate the implementation of different water management strategies. In order to meet goals or decisions set for water management, better coordination between different governmental sectors should be sought, and institutional and legal frameworks should be reviewed (Bates et al. 2008).

Further, proper public education and awareness of possible climate change impacts on water supply are also important to reduce water demand and pollution.

To reduce evaporation loss, which will increase due to global warming harvested rainwater may be placed underground in the sand-filled holes having impermeable geo-synthetic liners. Rainwater harvesting, stormwater and grey water recycling and solar distillation plants (Al-Sulttani et al. 2017) can assist to manage water resources at household levels, which will enhance the resilience of our water supply systems. The home shower systems should be timed to reduce shower water use. Rainwater tank sizing should be more realistic based on the local rainfall and water demand data to achieve increased water savings from the mains, the current 3 kL tank in most of the Sydney households is inadequate as found by A. Rahman, Keane, and Imteaz (2012). All the irrigation taps and the outdoor washing facilities must be connected to rainwater tanks only.

One of the important factor that could control water consumption is price, but the price should be sufficiently high to control the behavior since simple increases did not result in long term reduction in water demand.

CONCLUSION

Climate change results in an increased intensity of precipitation and melting of glaciers which affect seasonal flows, water temperature, etc. which in turn decline the surface and groundwater quality and add stress on water supplies. The impact of climate change and extreme weather events on water quality and availability are already being felt in communities around the world and future climate change is expected to further deteriorate the water quality and disrupt the many facets of life. Extreme weather and climate related impacts on water supply can result in increased risks in other two critical systems; food production and distribution, and energy production. As the future risks from climate change on water supply depend primarily on the decision made today, the chapter calls for better engineering and policy interventions. Therefore, water management strategies and coherent policy-making can help to reduce future risks and associated uncertainties in water supply due to climate change.

REFERENCES

Al-Sulttani, Ali Omran, Amimul Ahsan, Ataur Rahman, N. N. Nik Daud, and Syazwani Idrus. 2017. "Heat transfer coefficients and yield analysis of a double-slope solar still hybrid with rubber scrapers: an experimental and theoretical study." *Desalination* 407: 61-74.

Al Mamoon, Abdullah, and Ataur Rahman. 2017. "Rainfall in Qatar: Is it changing?" *Natural Hazards* 85 (1): 453-470.

Arnell, Nigel W., Sarah J. Halliday, Richard W. Battarbee, Richard A. Skeffington, and Andrew J Wade. 2015. "The implications of climate change for the water environment in England." *Progress in Physical Geography* 39 (1): 93-120.

Bates, B. C., Z. W. Kundzewicz, S. Wu, and J. P. Palutikof. 2008. *Climate Change and Water. Technical Paper VI of the Intergovernmental Panel on Climate Change.* IPCC Secretariat (Geneva, Switzerland).

Bianchi, Thomas S., Fenix Garcia-Tigreros, Shari A. Yvon-Lewis, Michael Shields, Heath J. Mills, David Butman, Christopher Osburn, Peter Raymond, G. Christopher Shank, Steven F. DiMarco, Nan Walker, Brandi Kiel Reese, Ruth Mullins-Perry, Antonietta Quigg, George R. Aiken, and Ethan L. Grossman. 2013. "Enhanced transfer of terrestrially derived carbon to the atmosphere in a flooding event." *Geophysical Research Letters* 40 (1): 116-122. https://doi.org/10.1029/2012gl054 145.

Bloomfield, J. P., R. J. Williams, D. C. Gooddy, J. N. Cape, and P. Guha. 2006. "Impacts of climate change on the fate and behaviour of pesticides in surface and groundwater—a UK perspective." *Science of The Total Environment* 369 (1): 163-177. https://doi.org/https://doi.org/10.1016/j.scitotenv.2006.05.019.

Chapagain, Ashok K., and Arjen Y. Hoekstra. 2008. "The global component of freshwater demand and supply: an assessment of virtual water flows between nations as a result of trade in agricultural and industrial products." *Water International* 33 (1): 19-32. https://doi.org/10.1080/02508060801927812.

Chiew, F. H. S., J. Teng, J. Vaze, D. A. Post, J. M. Perraud, D. G. C. Kirono, and N. R. Viney. 2009. "Estimating climate change impact on runoff across southeast Australia: Method, results, and implications of the modeling method." *Water Resources Research* 45 (10).

Delpla, I., A. V. Jung, E. Baures, M. Clement, and O. Thomas. 2009. "Impacts of climate change on surface water quality in relation to drinking water production." *Environment International* 35 (8): 1225-1233. https://doi.org/https://doi.org/10.1016/j.envint.2009.07.001.

Drewry, J. J., L. T. H. Newham, and B. F. W. Croke. 2009. "Suspended sediment, nitrogen and phosphorus concentrations and exports during storm-events to the Tuross estuary, Australia." *Journal of Environmental Management* 90 (2): 879-887. https://doi.org/https://doi.org/10.1016/j.jenvman.2008.02.004.

Evans, C. D., D. T. Monteith, and D. M. Cooper. 2005. "Long-term increases in surface water dissolved organic carbon: Observations, possible causes and environmental impacts." *Environmental Pollution* 137 (1): 55-71. https://doi.org/https://doi.org/10.1016/j.envpol.2004.12.031.

Fenner, N, C Freeman, Steven Hughes, and Brian Reynolds. 2001. "Molecular weight spectra of dissolved organic carbon in a rewetted Welsh peatland and possible implications for water quality." *Soil Use and Management* 17 (2): 106-112.

Funari, E., M. Manganelli, and L. Sinisi. 2012. "Impact of climate change on waterborne diseases." *Ann Ist Super Sanita* 48 (4): 473-87. https://doi.org/10.4415/ann_12_04_13.

Gaudard, Ludovic, Franco Romerio, Francesco Dalla Valle, Roberta Gorret, Stefano Maran, Giovanni Ravazzani, Markus Stoffel, and Michela Volonterio. 2014. "Climate change impacts on hydropower in the Swiss and Italian Alps." *Science of The Total Environment* 493: 1211-1221. https://doi.org/https://doi.org/10.1016/j.scitotenv.2013.10.012.

Gleick, Peter H., and Meena Palaniappan. 2010. "Peak water limits to freshwater withdrawal and use." *Proceedings of the National Academy of Sciences* 107 (25): 11155-11162.

Hajani, Evan, and Ataur Rahman. 2018. "Characterizing changes in rainfall: a case study for New South Wales, Australia." *International Journal of Climatology* 38 (3): 1452-1462.

Haque, Md. Mahmudul, Ataur Rahman, Dharma Hagare, Golam Kibria, and Fazlul Karim. 2015. "Estimation of catchment yield and associated uncertainties due to climate change in a mountainous catchment in Australia." *Hydrological Processes* 29 (19): 4339-4349.

Haque, Md. Mahmudul, Ataur Rahman, and Bijan Samali. 2016. "Evaluation of climate change impacts on rainwater harvesting." *Journal of Cleaner Production* 137: 60-69.

Hernando, M. D., M. Mezcua, A. R. Fernández-Alba, and D. Barceló. 2006. "Environmental risk assessment of pharmaceutical residues in wastewater effluents, surface waters and sediments." *Talanta* 69 (2): 334-342. https://doi.org/https://doi.org/10.1016/j.talanta.2005.09.037.

Hofstra, Nynke. 2011. "Quantifying the impact of climate change on enteric waterborne pathogen concentrations in surface water." *Current Opinion in Environmental Sustainability* 3 (6): 471-479. https://doi.org/https://doi.org/10.1016/j.cosust.2011.10.006.

Hossain, A. S. M., and A. Rahman. 2019. "Trend Analysis in Flood Data in the Brisbane River Catchment, Australia." 2nd International Conference on Water and Environmental Engineering (iCWEE-2019), Dhaka, Bangladesh, 19-22 January 2019.

Hounshell, Alexandria G., Jacob C. Rudolph, Bryce R. Van Dam, Nathan S. Hall, Christopher L. Osburn, and Hans W. Paerl. 2019. "Extreme weather events modulate processing and export of dissolved organic carbon in the Neuse River Estuary, NC." *Estuarine, Coastal and Shelf*

Science 219: 189-200. https://doi.org/https://doi.org/10.1016/j.ecss. 2019.01.020.

Hua, Guanghui, and David A. Reckhow. 2008. "DBP formation during chlorination and chloramination: Effect of reaction time, pH, dosage, and temperature." *American Water Works Association. Journal* 100 (8): 82-95,12.

Hunter, P. R. 2003. "Climate change and waterborne and vector-borne disease." *Journal of Applied Microbiology* 94 (s1): 37-46. https://doi.org/10.1046/j.1365-2672.94.s1.5.x.

Huo, Shouliang, Zhuoshi He, Chunzi Ma, Hanxiao Zhang, Beidou Xi, Xinghui Xia, Ying Xu, and Fengchang Wu. 2019. "Stricter nutrient criteria are required to mitigate the impact of climate change on harmful cyanobacterial blooms." *Journal of Hydrology* 569: 698-704. https://doi.org/ https://doi.org/10.1016/j.jhydrol.2018.12.030.

International Energy Agency (IEA). 2012. *World Energy Outlook 2012: Executive Summary.* (Paris, France). https://www.iea.org/publications/ freepublications/publication/English.pdf.

Ishak, Elias H., Ataur Rahman, Seth Westra, Ashish Sharma, and George Kuczera. 2013. "Evaluating the non-stationarity of Australian annual maximum flood." *Journal of Hydrology* 494: 134-145.

Janssen, E., R. L. Sriver, D. J. Wuebbles, and K. E. Kunkel. 2016. "Seasonal and regional variations in extreme precipitation event frequency using CMIP5." *Geophysical Research Letters* 43 (10): 5385-5393. https://doi.org/10.1002/2016gl069151.

Jeppesen, Erik, Brian Kronvang, Mariana Meerhoff, Martin Søndergaard, Kristina M Hansen, Hans E Andersen, Torben L Lauridsen, Lone Liboriussen, Meryem Beklioglu, and Arda Özen. 2009. "Climate change effects on runoff, catchment phosphorus loading and lake ecological state, and potential adaptations." *Journal of Environmental Quality* 38 (5): 1930-1941.

Kastl, G., A. Sathasivan, and I. Fisher. 2016. "A selection framework for NOM removal process for drinking water treatment." *Desalination and Water Treatment* 57 (17): 7679-7689. https://doi.org/10.1080/ 19443994.2015.1044476.

Kastl, George, Arumugam Sathasivan, I. A. N. Fisher, and John Van Leeuwen. 2004. "Modeling DOC Removal by Enhanced Coagulation." *Journal of American Water Works Association* 96 (2): 79-89. https://doi.org/10.2307/41312060.

Korotta-Gamage, Shashika Madushi, and Arumugam Sathasivan. 2017. "A review: Potential and challenges of biologically activated carbon to remove natural organic matter in drinking water purification process." *Chemosphere* 167: 120-138. https://doi.org/http://dx.doi.org/10.1016/j.chemosphere.2016.09.097.

Kosten, Sarian, Vera L. M. Huszar, Eloy Bécares, Luciana S. Costa, Ellen Van Donk, Lars-Anders Hansson, Erik Jeppesen, Carla Kruk, Gissell Lacerot, and Néstor Mazzeo. 2012. "Warmer climates boost cyanobacterial dominance in shallow lakes." *Global Change Biology* 18 (1): 118-126.

Laz, Orpita U., Ataur Rahman, Abdullah Yilmaz, and Khaled Haddad. 2014. "Trends in sub-hourly, sub-daily and daily extreme rainfall events in eastern Australia." *Journal of Water and Climate Change* 5 (4): 667-675.

Lipczynska-Kochany, Ewa. 2018. "Effect of climate change on humic substances and associated impacts on the quality of surface water and groundwater: A review." *Science of The Total Environment* 640-641: 1548-1565. https://doi.org/https://doi.org/10.1016/j.scitotenv.2018.05.376.

Liu, Xingcai, Qiuhong Tang, Nathalie Voisin, and Huijuan Cui. 2016. "Projected impacts of climate change on hydropower potential in China." *Hydrology and Earth System Sciences* 20 (8): 3343-3359. https://doi.org/http://dx.doi.org/10.5194/hess-20-3343-2016.

Loo, Siew-Leng, Anthony G. Fane, William B. Krantz, and Teik-Thye Lim. 2012. "Emergency water supply: A review of potential technologies and selection criteria." *Water Research* 46 (10): 3125-3151. https://doi.org/https://doi.org/10.1016/j.watres.2012.03.030.

Magazinovic, Rodney S., Brenton C. Nicholson, Dennis E. Mulcahy, and David E. Davey. 2004. "Bromide levels in natural waters: its relationship to levels of both chloride and total dissolved solids and the implications for water treatment." *Chemosphere* 57 (4): 329-335. https://doi.org/https://doi.org/10.1016/j.chemosphere.2004.04.056.

Majone, Bruno, Francesca Villa, Roberto Deidda, and Alberto Bellin. 2016. "Impact of climate change and water use policies on hydropower potential in the south-eastern Alpine region." *Science of The Total Environment* 543: 965-980. https://doi.org/https://doi.org/10.1016/j.scitotenv.2015.05.009.

Masiá, Ana, Julián Campo, Pablo Vázquez-Roig, Cristina Blasco, and Yolanda Picó. 2013. "Screening of currently used pesticides in water, sediments and biota of the Guadalquivir River Basin (Spain)." *Journal of Hazardous Materials* 263: 95-104. https://doi.org/https://doi.org/10.1016/j.jhazmat.2013.09.035.

Me, Wang, David P. Hamilton, Christopher G. McBride, Jonathan M. Abell, and Brendan J. Hicks. 2018. "Modelling hydrology and water quality in a mixed land use catchment and eutrophic lake: Effects of nutrient load reductions and climate change." *Environmental Modelling & Software* 109: 114-133. https://doi.org/https://doi.org/10.1016/j.envsoft.2018.08.001.

Mohammed, Hadi, Ann-Kristin Tveten, and Razak Seidu. 2019. "Modelling the impact of climate change on flow and E. coli concentration in the catchment of an ungauged drinking water source in Norway." *Journal of Hydrology* 573: 676-687. https://doi.org/ https://doi.org/10.1016/j.jhydrol.2019.04.021.

National Academies of Sciences, Engineering, and Medicine. 2019. *Environmental Engineering for the 21st Century: Addressing Grand Challenges*. Washington, DC: The National Academies Press.

Osorio, Victoria, Rafael Marcé, Sandra Pérez, Antoni Ginebreda, Jose Luís Cortina, and Damià Barceló. 2012. "Occurrence and modeling of pharmaceuticals on a sewage-impacted Mediterranean river and their dynamics under different hydrological conditions." *Science of The Total Environment* 440: 3-13. https://doi.org/ https://doi.org/10.1016/j.sci totenv.2012.08.040.

Ouarda, Taha B. M. J., and Christian Charron. 2018. "Nonstationary Temperature-Duration-Frequency curves." *Scientific Reports* 8 (1): 15493. https://doi.org/10.1038/s41598-018-33974-y.

Pagano, Todd, Morgan Bida, and Jonathan E. Kenny. 2014. "Trends in Levels of Allochthonous Dissolved Organic Carbon in Natural Water: A Review of Potential Mechanisms under a Changing Climate." *Water* 6 (10): 2862-2897. https://doi.org/http://dx.doi.org/10.3390/w6102862.

Pivokonsky, Martin, Jana Naceradska, Tomas Brabenec, Katerina Novotna, Magdalena Baresova, and Vaclav Janda. 2015. "The impact of interactions between algal organic matter and humic substances on coagulation." *Water Research* 84: 278-285. https://doi.org/https://doi.org/10.1016/j.watres.2015.07.047.

Rahman, Ataur, Joseph Keane, and Monzur Alam Imteaz. 2012. "Rainwater harvesting in Greater Sydney: Water savings, reliability and economic benefits." *Resources, Conservation and Recycling* 61: 16-21. https://doi.org/https://doi.org/10.1016/j.resconrec.2011.12.002.

Rahman, M. M., M. A. Rahman, M. M. Haque, and A. Rahman. 2019. *Sustainable Water Use in Construction*. Edited by Vivian Y. Tam and Khoa N Le. *Sustainable Construction Technologies: Life-cycle Assessment*: Butterworth-Heinemann.

Rakib, M. A., Jun Sasaki, Hirotaka Matsuda, and Mayumi Fukunaga. 2019. "Severe salinity contamination in drinking water and associated human health hazards increase migration risk in the southwestern coastal part of Bangladesh." *Journal of Environmental Management* 240: 238-248. https://doi.org/https://doi.org/10.1016/j.jenvman.2019.03.101.

Raymond, Peter A., and James E. Saiers. 2010. "Event controlled DOC export from forested watersheds." *Biogeochemistry* 100 (1-3): 197-209.

Raymond, Peter A., James E. Saiers, and William V. Sobczak. 2016. "Hydrological and biogeochemical controls on watershed dissolved organic matter transport: Pulse-shunt concept." *Ecology* 97 (1): 5-16.

Ritson, J. P., N. J. D. Graham, M. R. Templeton, J. M. Clark, R. Gough, and C. Freeman. 2014. "The impact of climate change on the treatability of dissolved organic matter (DOM) in upland water supplies: A UK perspective." *Science of The Total Environment* 473-474: 714-730. https://doi.org/https://doi.org/10.1016/j.scitotenv.2013.12.095.

Schmeller, Dirk S., Adeline Loyau, Kunshan Bao, Werner Brack, Antonis Chatzinotas, Francois De Vleeschouwer, Jan Friesen, Laure Gandois, Sophia V. Hansson, Marilen Haver, Gaël Le Roux, Ji Shen, Roman Teisserenc, and Vance T. Vredenburg. 2018. "People, pollution and pathogens – Global change impacts in mountain freshwater ecosystems." *Science of The Total Environment* 622-623: 756-763. https://doi.org/https://doi.org/10.1016/j.scitotenv.2017.12.006.

Sharma, Ashish, Conrad Wasko, and Dennis P. Lettenmaier. 2018. "If Precipitation Extremes Are Increasing, Why Aren't Floods?" *Water Resources Research* 54 (11): 8545-8551. https://doi.org/10.1029/2018wr023749.

Stellar, D. 2010. "Can we have our water and drink it, too? Exploring the water quality-quantity nexus." State of the Planet blog, New York, Earth Institute, Columbia University. Accessed 2019 April. https://blogs.ei.columbia.edu/2010/10/28/can-we-have-our-water-and-drink-it-too-exploring-the-water-quality-quantity-nexus/.

Tang, R., J. M. Clark, T. Bond, N. Graham, D. Hughes, and C. Freeman. 2013. "Assessment of potential climate change impacts on peatland dissolved organic carbon release and drinking water treatment from laboratory experiments." *Environmental Pollution* 173: 270-277. https://doi.org/https://doi.org/10.1016/j.envpol.2012.09.022.

Thies, Hansjörg, Ulrike Nickus, Volkmar Mair, Richard Tessadri, Danilo Tait, Bertha Thaler, and Roland Psenner. 2007. "Unexpected response of high alpine lake waters to climate warming." *Environmental science & technology* 41 (21): 7424-7429.

Trolle, Dennis, David P. Hamilton, Conrad A. Pilditch, Ian C. Duggan, and Erik Jeppesen. 2011. "Predicting the effects of climate change on trophic status of three morphologically varying lakes: Implications for lake restoration and management." *Environmental Modelling & Software* 26 (4): 354-370. https://doi.org/https://doi.org/10.1016/j.envsoft.2010.08.009.

Turral, Hugh, Jacob Burke, and Jean-Marc Faurès. 2011. *Climate change, water and food security.* Food and Agriculture Organization of the United Nations (Rome, Italy).

U.S. Global Change Research Program (USGCRP). 2018. *Impacts, Risks, and Adaptation in the United States: Fourth National Climate Assessment, Volume II.* U.S. Government Publishing Office (Washington, DC).

UN Water. 2018. *Sustainable Development Goal 6: Synthesis Report 2018 on Water and Sanitation.* (New York: United Nations Publications).

United Nations. 2012. *WWAP (World Water Assessment Programme). The United Nations World Water Development Report 4: Managing Water under Uncertainty and Risk.* (Paris UNESCO).

____. 2018. *The Sustainable Development Goals Report.* (New York: United Nations Publications).

United States Environmental Protection Agency (USEPA). 2017. "Climate Impacts on Agriculture and Food Supply." https://19january2017 snapshot.epa.gov/climate-impacts/climate-impacts-agriculture-and-food-supply_.html.

van Vliet, M. T. H., and J. J. G. Zwolsman. 2008. "Impact of summer droughts on the water quality of the Meuse river." *Journal of Hydrology* 353 (1): 1-17. https://doi.org/https://doi.org/10.1016/j.jhydrol.2008.01.001.

Watts, Carol D., Pamela S. Naden, John Machell, and Jenny Banks. 2001. "Long term variation in water colour from Yorkshire catchments." *Science of The Total Environment* 278 (1): 57-72. https://doi.org/https://doi.org/10.1016/S0048-9697(00)00888-3.

Whitehead, P. G., R. L. Wilby, R. W. Battarbee, M. Kernan, and A. J. Wade. 2009. "A review of the potential impacts of climate change on surface water quality." *Hydrological Sciences Journal* 54 (1): 101-123. https://doi.org/10.1623/hysj.54.1.101.

Winid, Bogumiła. 2015. "Bromine and water quality – Selected aspects and future perspectives." *Applied Geochemistry* 63: 413-435. https://doi.org/https://doi.org/10.1016/j.apgeochem.2015.10.004.

World Health Organization (WHO). 2016. *Preventing disease through healthy environments: a global assessment of the burden of disease from environmental risks* (France).

Worrall, Fred, Ron Harriman, Chris D. Evans, Carol D. Watts, John Adamson, Colin Neal, Ed Tipping, Tim Burt, Ian Grieve, Don Monteith, Pam S. Naden, Tom Nisbet, Brian Reynolds, and Paul Stevens. 2004. "Trends in Dissolved Organic Carbon in UK Rivers and Lakes." *Biogeochemistry* 70 (3): 369-402. https://doi.org/10.1007/s10533-004-8131-7.

Wright, Benjamen, Benjamin D. Stanford, Allison Reinert, Jan C. Routt, Stuart J. Khan, and Jean F. Debroux. 2014. "Managing water quality impacts from drought on drinking water supplies." *Journal of Water Supply: Research and Technology - AQUA* 63 (3): 179-188. https://doi.org/http://dx.doi.org/10.2166/aqua.2013.123.

WWAP (United Nations World Water Assessment Programme). 2014. *The United Nations World Water Development Report 2014: Water and Energy*. UNESCO (Paris).

Zwolsman, J. J. G., and A. J. Van Bokhoven. 2007. "Impact of summer droughts on water quality of the Rhine River-a preview of climate change?" *Water Science and Technology* 56 (4): 45-55.

In: Sustainable Water
Editors: V. Gnaneswar Gude et al.
ISBN: 978-1-53617-335-2

Chapter 6

BUFFALO FORAGE AND WATER ESTIMATION: MANAGEMENT DECISIONS AND ASSISTING CLIMATE CHANGE VULNERABILITY ASSESSMENTS AND DROUGHT MANAGEMENT ADAPTATION

Joseph M. Wilder[1], Kurt M. Chowanski[2], PhD,
Scott J. Kenner[1], PhD
and Venkata R. Gadhamshetty[1,*], PhD
[1]Department of Environmental and Civil Engineering
South Dakota School of Mines and Technology,
Rapid City, South Dakota, US
[2]Department of Chemistry and Applied Biological Sciences,
South Dakota School of Mines and Technology,
Rapid City, South Dakota, US

---

* Corresponding Author's Email: venkata.gadhamshetty@sdsmt.edu.

ABSTRACT

This chapter details application of the developed Buffalo Forage and Water Estimator Tools. Together, these tools estimate a pasture's total annual forage production and combines it with user-specified herd size, gender ratio, and desired pasture utilization to analyze pasture impact, need for supplemental feeding, and water requirements. The tools provide essential information to help Tribes make informed buffalo management decisions and enables adaptive management as it estimates the effects of preparedness or response adaptations (e.g., reducing herd size during a period of drought). Application of these tools is demonstrated for two pilot projects: Taos Pueblo of New Mexico and the Ponca Tribe of Nebraska. For Taos Pueblo, these tools estimated the number of buffalo that their pasture can sustain annually during unfavorable weather conditions as well as the amount of supplemental forage required to sustain their desired herd size. For the Ponca Tribe, these tools estimated the amount of additional forage required to sustainably maintain their current herd structure as well as the water requirements, which significantly influenced water treatment considerations.

Keywords: climate change, sustainability, resiliency, drought, water treatment, tribe, buffalo, adaptation, adaptive capacity, vulnerability assessment

INTRODUCTION

This book chapter summarizes the author's thesis work (Wilder, 2018), which was a continuation of a project done by South Dakota School of Mines and Technology in collaboration with the InterTribal Buffalo Council. This chapter includes parts of the author's contribution from both projects, often verbatim.

The Buffalo Forage and Water Estimator Tools discussed in this chapter were developed to inform management decisions and assist climate change vulnerability assessments and drought management adaptation. This included providing guidance and recommendations to Tribes on how to manage their available forage and water. Forage management included estimating how much forage would be available and how many buffalo it could support. Water management included estimating how much water would be required for buffalo based on the forage availability and suggesting alternatives or best management practices.

The ArcMap-based Forage Estimator Model tool estimates a pasture's annual forage production for three growing season conditions based on spatially explicit estimates of forage productivity by soil type from the Natural Resources Conservation Service (NRCS) Web Soil Survey. The spreadsheet-based Forage and Water Estimator Tool combines the annual forage production estimates from the Forage Estimator Model with user-specified inputs of herd size, gender ratio, and desired pasture utilization to analyze the pasture impact, need for supplemental feeding, and water requirements. The Forage and Water Estimator Tool is interactive and enables adaptive management as it estimates the effects of preparedness or response adaptations. The usefulness of these tools is illustrated by application in two pilot projects.

General Information on Buffalo Management

Buffalo (*Bison bison)* are a culturally important species to Native Americans and are managed for various purposes such as promoting spiritual revitalization, cultural enhancement, ecological restoration, and economic development. The term buffalo is used in accordance with Tribal preferences. Buffalo can be used as a source of meat for Tribal members and herds are often managed for other cultural uses. Sustainable management is essential to maintain Tribal buffalo herds for future generations. A forage availability assessment was conducted for the Taos Pueblo and Ponca Tribes to understand how future conditions may impact their buffalo herds. Insight about future conditions can be used to help make buffalo management decisions. A goal of this project was to provide guidance and recommendations to Tribes pertaining to management of forage and water for buffalo.

Understanding how resources may change in the future may help Tribal resource managers maintain productive lands and water availability.

Understanding the supply and demand of forage and water of a buffalo pasture is essential to making informed herd management decisions. Although both forage and water are essential to maintaining buffalo herds, it has been assumed that impacts from climate change will affect the availability of forage more than the availability of water (McNeeley et al. 2016, Derner et al. 2015, Polley et al. 2013, Nippert et al. 2013, National Drought Mitigation Center 2012, Duesterhaus et al. 2008). This is a rather large assumption as the availability of water for buffalo has seldomly been evaluated in a changing climate. Historical data is used to estimate future variations for the availability of water and is the basis for current policies and decision-making tools. Unfortunately, the water sources used by livestock such as cattle or buffalo are not addressed frequently. In a changing climate, forage and water resources are likely to be affected; this may complicate buffalo management. By providing the tools necessary to understand current forage and water resources, Tribal resource managers should be able to plan and make more informed decisions. Insight from predictive climate change models provide knowledge necessary to plan for a changing climate, which can be applied in the Forage and Water Estimator Tool.

InterTribal Buffalo Council Climate Change Vulnerability Assessment and Effective Drought Management Adaptation report (2017) assessed forage and water resources with a comprehensive compilation of observed and projected data. The observed data was used to help understand current resources and the projected data was used to help understand how these resources may change in the future. Impacts from climate change may complicate buffalo management. Therefore, impacts from climate change, such as reduced forage availability, can be applied in the Forage and Water Estimator Tool to see how they may affect the amount of buffalo that can be supported. Using this tool, one can estimate the effects of preparedness or response adaptations which may reduce the impacts of climate change. The goal of developing this tool was to help Tribal resource managers make informed buffalo management decisions. The objectives of this research were to (1) develop a tool that can help Tribes understand availability and requirement of forage and water for buffalo; (2) use the developed tool to help inform buffalo management, such as the number of buffalo that can be supported by a pasture; (3) use the developed tool to help inform water requirements; (4) use the developed tool to improve future outcomes of management decisions by increasing knowledge and capacity for sustainable and adaptive buffalo management; (5) use the developed tool to help determine the level of vulnerability and estimate the effects of preparedness or response adaptations.

LITERATURE REVIEW

Although there are significant differences between buffalo and cattle, management of the two has generally been considered equivalent. Literature that provides specific guidelines for buffalo management were limited but were incorporated when available. This study focused on the availability and requirement of forage and water, as both are likely to be impacted by climate change.

Forage Demand for Buffalo Management

Accurately estimating the amount of forage that a pasture will produce is critical to sustainable management of buffalo herds. By estimating forage availability, one can characterize the risk of overgrazing. Overgrazing, or utilizing more than 50% of annual forage production, should be avoided as it may result in negative consequences, such as reduced subsequent-year forage production (University of Nebraska 2018). Additional forage should be provided if buffalo are utilizing more than 50% of the total annual forage production. Forage availability can be increased by providing hay or increasing pasture area; or forage requirement can be reduced by decreasing the number of buffalo.

Forage availability should be estimated for pastures individually, as soil type directly affects forage productivity. Therefore, total annual forage estimated for one pasture is likely to be different than another pasture, even if it is the same exact size. Total annual forage production can be estimated based on spatially explicit estimates of forage productivity by soil type for rangeland soils, which are provided by the NRCS Web Soil Survey (Soil Survey Staff et al.). The parameters used to define favorable, normal, and unfavorable growing conditions are not defined, although they are likely based on field data, precipitation, temperature, soil moisture, et cetera. Favorable growing conditions will yield the highest forage production and unfavorable growing conditions will yield the lowest forage production. Total forage production for a given pasture can be estimated by forage productivity estimates and the area of each soil type. For the purposes of this project, growing season favorability was characterized by the Palmer Drought Severity Index.

Total annual forage production that is estimated by the Forage and Water Estimator Tool represents an estimated theoretical maximum. It is likely that actual forage availability may be significantly lower than the estimated value, as (1) some forage may not be desirable or edible to the buffalo; (2) some forage may be fouled by trampling and defecating; (3) other animals may be grazing; (4) some forage has to be left on the flora so that they keep growing. Holechek et al. (1989) uses the phrase "take half, leave half", which seems to be a sustainable grazing practice which can maintain forage production. However, recent studies suggest that 50% defoliation resulted in reduced subsequent-year forage production by around 20% (University of Nebraska 2018). Therefore, it seems inadvisable to manage at 50% of total forage production. Additionally, as precipitation is so variable, managing at 50% of total forage production can result in a much higher allocation during a year with lower than normal precipitation. A total forage allocation of 50% during a favorable year may be equal to 100% allocation during an unfavorable year. If this was to occur, either supplement feeding or decreasing herd size would be required. Adjusting herd size each year based on actual or predicted growing season conditions is infeasible and may be expensive. One of the main objectives of many tribes maintaining buffalo is to provide food; having a stable herd size may be more desirable than adjusting every year.

If herd size is based on forage production estimates for unfavorable growing seasons only, then the need for herd size intervention decreases. Although managing herd size based on unfavorable growing conditions limits buffalo production, it will likely sustain long-term protection over grazing resources. A herd size that yields consumption to around 25% of annual forage production is suggested and considered sustainable (Holechek, 1988) (Holechek & Pieper, 1992). At 25% allocation, it is unlikely that annual herd adjustments will be required.

Animal unit time is a convenient way to compare forage allocations. Animal unit time include animal unit days (AUD), animal unit months (AUM), and animal unit years (AUY). Bedell (1998) notes that the Society of Range Management considers 1 animal unit to be a mature cow weighing about 1,000 pounds with or without a calf of up to six months old. Equivalently, any animal consuming around 26 pounds of forage per day is considered to be 1 animal unit (Bedell, 1998). As a male buffalo (bull) consumes more forage, they are considered to be 1.5 animal units (United States Department of Agriculture and Natural Resources Conservation Service, 1997). One animal unit day is the amount of forage required by one animal unit for one day, on an oven-dry basis (Bedell, 1998). Likewise, AUM and AUY is the amount of forage required by one animal unit for one month and one year, respectively. The actual forage requirement for buffalo has not been well documented, therefore, a value of 30 pounds per AUD is assumed for the use of this project.

The Forage and Water Estimator Tool provides a range of forage allocation percentages that can be allocated to buffalo. An upper allocation limit of 50% of the total annual forage production was used. Again, although there is some evidence suggesting this level of defoliation is damaging (University of Nebraska, 2018), this is the generally accepted maximum (Holechek et al. 1989).

Lack of documentation or formal studies quantifying buffalo forage consumption results in some uncertainty in the estimates of annual forage requirement. It may be beneficial to measure buffalo forage consumption in a controlled environment, in which forage is provided and buffalo do not have access to natural food sources.

Water Demand for Buffalo Management

A definitive amount of water that buffalo consume has not been documented (Nippert et al. 2013). However, it is known that buffalo require the highest demand for water during the hottest days of the year. Several studies and personal estimates suggest that a buffalo will require from 12 to 30 gallons of water per day. One study done in Alberta, Canada, suggests that the annual average demand is 12 gallons per day (Alberta Agriculture and Forestry, 2009). This estimate is likely a reasonable annual average for the respective climate; however, it is probably substantially lower than what buffalo consume during the hottest days of the year, or in hotter climates. Kohl et al. (2013) notes that buffalo increase the amount of time they spend at water sources as temperature increases. This implies that buffalo increase their water consumption in response to increased temperatures. Winchester and Morris (1956) reported that cattle's water consumption at temperatures above 90°F increase by a factor of 1.255. If the same rate of increase in water consumption for buffalo is assumed and assuming 12 gallons per day is the annual average, then a peak water demand for buffalo is estimated to be 15 gallons per day during the hottest days of the year.

Nippert et al. (2013) found that buffalo get between 19% and 25% of their water consumption from plant tissue. Based on unpublished data from T. Ocheltree, Nippert et al. (2013) suggests that upwards of 98% of grasses weight is water, when soil moisture is readily available. If temperatures increase or precipitation decreases, or evaporation increases more than the increase in precipitation, it is likely that plant tissue water will be reduced. As a result, buffalo may require additional drinking water.

Based on 25 years of experience, Chad Kremer, Custer State Park Buffalo Herd Manager, advised that a reasonable estimate of buffalo water requirement in the summer is upwards of 20 gallons per day (Chad Kremer, email communication, February 28, 2017). Likewise, Moritz Espy, ranch manager of the 777 Buffalo Ranch in Hermosa, South Dakota, estimated that buffalo's peak water demand during the hottest part of the year is between 20-30 gallons per day (Moritz Espy, personal communication, February 28, 2017).

Lack of documentation or formal studies quantifying buffalo water consumption results in some uncertainty in the estimates of daily water requirement.

It may be beneficial to measure buffalo water consumption in a controlled environment, in which water is provided and buffalo do not have access to natural water sources.

DEVELOPED TOOLS

The methods used to develop the tools are not included in this chapter. The final products of this research project were an ArcMap ModelBuilder model called the Forage Estimator Tool and an interactive Microsoft Excel workbook called the Automated Forage and Water Estimator Tool. The Forage Estimator Tool uses NRCS Web Soil Survey data on ArcMap to estimate annual forage production for three different growing conditions for a given buffalo pasture.

The Automated Forage and Water Estimator Tool uses the forage production values from the Forage Estimator Tool to determine how many buffalo the pasture can support, how much water the buffalo will require, and how much supplemental forage, if any, should be provided. Together, these tools can help Tribes manage their buffalo herds and help them estimate the effects of making decisions regarding adaptive management practices. Originally, for the pilot projects, the non-automated version of the Forage and Water Estimator Tool was used.

RESULTS

This section shows the potential usefulness of the developed tools. First, two pilot projects are presented to illustrate how the Buffalo Forage and Water Estimator Tool estimates forage and water requirements and how that can be used inform management decisions, water development options, climate change vulnerability assessments, and recommendations. The second part shows how the automated version simplifies the process.

Taos Pueblo, New Mexico

Introduction

Taos Pueblo buffalo managers were interested in developing a new pasture for their buffalo. The Tribe's buffalo managers were interested in assessing food and water resources and developing a water source for the pasture, so they can maintain a herd of 50-70 buffalo. The proposed buffalo pasture had been grazed by cattle in the past and contains one abandoned well. Development of a permanent and reliable water source is necessary to allow sustainable buffalo management. The proposed buffalo pasture consists of around 4,900 acres of land. The goal of this pilot project was to assess forage production within the buffalo pasture, estimate how many buffalo can be supported annually, and provide examples of water development options.

Forage and Water Supply and Demand

Total annual forage production, animal unit time, herd size, and water demand for the Buffalo Pasture were estimated with the Forage and Water Estimator Tool based on spatial data provided by Taos Pueblo (email communication, April-May 2017) and the NRCS Web Soil Survey (Soil Survey Staff et al.). Forage productivity by soil type from the Web Soil Survey was copied and pasted into the first worksheet (Figure 1).

Forage production is calculated from the soil productivity estimates. The total pasture forage production is calculated for each growing condition (Figure 2).

Annual forage production of the pasture ranges from 1,931,965 – 3,761,696 pounds depending on the favorability of the growing season. Figure 3 shows animal unit time for a range of forage allocations calculated by the Forage and Water Estimator Tool.

Worksheet 1- This worksheet is where the soil productivity data is entered

Web Soil Survey (NRCS) Range Soil Production Estimates for favorable, normal, and unfavorable growing seasons

Favorable Productivity

Summary by Map Unit — Taos County and Parts of Rio Arriba and Mora Counties, New Mexico (NM670)

Map unit symbol	Map unit name	Rating (pounds per acre per year)	Acres in AOI	Percent of AOI
FbC	Fernando silt loam, 0 to 7 percent slopes	970	1,930.4	45.0%
PfC	Petaca-Prieta complex, 1 to 8 percent slopes	595	582.2	13.6%

Normal Productivity

Summary by Map Unit — Taos County and Parts of Rio Arriba and Mora Counties, New Mexico (NM670)

Map unit symbol	Map unit name	Rating (pounds per acre per year)	Acres in AOI	Percent of AOI
FbC	Fernando silt loam, 0 to 7 percent slopes	619	1,930.4	45.0%
PfC	Petaca-Prieta complex, 1 to 8 percent slopes	281	582.2	13.6%

Unfavorable Productivity

Summary by Map Unit — Taos County and Parts of Rio Arriba and Mora Counties, New Mexico (NM670)

Map unit symbol	Map unit name	Rating (pounds per acre per year)	Acres in AOI	Percent of AOI
FbC	Fernando silt loam, 0 to 7 percent slopes	513	1,930.4	45.0%
PfC	Petaca-Prieta complex, 1 to 8 percent slopes	270	582.2	13.6%

Figure 1. Forage and Water Estimator Tool customized for Taos Pueblo, worksheet 1.

Worksheet 2- This worksheet is where the total pasture productivity is calculated

This table gets the Web Soil Survey data for favorable normal and unfavorable conditions into one place and calculates the total pasture production. The productivity for each soil type is multiplied by the area in acres to get a production value for each soil type. The production values for all soil types in the pasture are then summed to get total pasture forage production.

		Forage Production (pounds)			*Area*
Map unit symbol	***Map unit name***	***Favorable***	***Normal***	***Unfavorable***	***(acres)***
FbC	Fernando silt loam, 0 to 7 percent slopes	1872488	1194918	990295	1,930.40
MNC	Manzano clay loam, 0 to 5 percent slopes	2970	0	1620	2.70
PfC	Petaca-Prieta complex, 1 to 8 percent slopes	346409	163598	157194	582.20
RcG	Rock outcrop, very steep	0	0	0	14.50
SED	Sedillo-Silva association, strongly sloping	19710	12960	10125	27.00
SgC	Servilleta-Prieta complex, 1 to 5 percent slopes	1152622	633735	583921	1,383.70
SSC	Silva-Sedillo association, gently sloping	367497	200326	188809	349.00
					Total Area
	Total Pasture Forage Production (pounds)	**3761696**	**2205536**	**1931965**	**4,289.50**

Figure 2. Forage and Water Estimator Tool customized for Taos Pueblo, worksheet 2.

Worksheet 3- This worksheet is where the pasture stocking capacity is estimated interms of Animal Unit Time

This set of tables estimate animal unit time that can be supported by the expected annual forage production estimate. Included is a range of amounts of forage that could be allocated to buffalo with an upper limit of allocating half of the annual total pasture forage production to buffalo; this level of forage consumption is a generally accepted maximum. Allowed annimal unit time is given in animal unit days (AUD), animal unit months (AUM), and animal unit years (AUY).

		Low Risk		Moderate Risk		High Risk
		Animal Unit Days (AUD)				
Growing Season Condition	**Total Pasture Forage Production**	**10% of forage allocated to**	**20% of forage allocated to**	**30% of forage allocated to**	**40% of forage allocated to**	**50% of forage allocated to**
Unfavorable	1931965	6440	12880	19320	25760	32199
Normal	2205536	7352	14704	22055	29407	36759
Favorable	3761696	12539	25078	37617	50156	62695

		Animal Unit Months (AUM)				
Growing Season Condition	**Total Pasture Forage Production**	**10% of forage allocated to**	**20% of forage allocated to**	**30% of forage allocated to**	**40% of forage allocated to**	**50% of forage allocated to**
Unfavorable	1931965	211	422	633	845	1056
Normal	2205536	241	482	723	964	1205
Favorable	3761696	411	822	1233	1644	2056

		Animal Unit Years (AUY)				
Growing Season Condition	**Total Pasture Forage Production**	**10% of forage allocated to**	**20% of forage allocated to**	**30% of forage allocated to**	**40% of forage allocated to**	**50% of forage allocated to**
Unfavorable	1931965	18	35	53	71	88
Normal	2205536	20	40	60	81	101
Favorable	3761696	34	69	103	137	172

Figure 3. Forage and Water Estimator Tool customized for Taos Pueblo, worksheet 3.

This worksheet represents the amount of Buffalo over a 1 year period

Steps 1&2 for PERCENTAGE RATIO of male and female buffalo
If the exact amount of desired male and female buffalo are known, use Steps 3 and 4

Step 1	Insert percentage of male buffalo desired in box B8
Step 2	Insert percentage of female buffalo desired in box C8

Gender	Male Buffalo	Female Buffalo	Total
Percentage of Buffalo	15%	85%	100%

The following three tables provide the amount of male/female Buffalo that can be supported based on available forage and herd structure. It also shows the water requirement.

Low Risk · Moderate Risk · High Risk

Unfavorable Growing Season					
% Forage Allocated	10%	20%	30%	40%	50%
Maximum Animal Units	18	35	53	71	88
Number of Buffalo	2 male 13 female	4 male 27 female	7 male 41 female	9 male 55 female	12 male 69 female
Water Requirement (gal/day)					
Low (15 gal)	225	465	720	960	1215
Average (20 gal)	300	620	960	1280	1620
High (30 gal)	450	930	1440	1920	2430

Normal Growing Season					
% Forage Allocated	10%	20%	30%	40%	50%
Maximum Animal Units	20	40	60	81	101
Number of Buffalo	2 male 15 female	5 male 31 female	8 male 47 female	11 male 63 female	14 male 79 female
Water Requirement (gal/day)					
Low (15 gal)	255	540	825	1110	1395
Average (20 gal)	340	720	1100	1480	1860
High (30 gal)	510	1080	1650	2220	2790

Favorable Growing Season					
% Forage Allocated	10%	20%	30%	40%	50%
Maximum Animal Units	34	69	103	137	172
Number of Buffalo	4 male 27 female	9 male 54 female	14 male 81 female	19 male 108 female	23 male 135 female
Water Requirement (gal/day)					
Low (15 gal)	465	945	1425	1905	2370
Average (20 gal)	620	1260	1900	2540	3160
High (30 gal)	930	1890	2850	3810	4740

Figure 4. Left side of worksheet 4 of the Forage and Water Estimator Tool customized for Taos Pueblo.

This pasture is estimated to support 18-172 AUY, depending on growing season favorability and the percentage of annual forage production allocated. The range of forage allocations presented represents a range of management choices that are consistent with generally accepted grazing management practices.

Risk of overgrazing is low when buffalo are allocated 10-25% of the total annual forage production for an unfavorable growing season, moderate at 25-40% of the total annual forage production, and high above 40% of the total annual forage production. When buffalo have consumed the allocated amount of forage, they should be fed hay or moved.

The number of buffalo that the pasture could support on an annual basis across a range of forage allocations was estimated for a herd of 15% males and 85% females. Figure 4 and Figure 5 shows AUY and the number of buffalo for a range of forage allocations calculated by the Forage and Water Estimator tool. It also provides an estimate of the amount of additional forage necessary for 53 buffalo.

This worksheet represents the amount of Buffalo over a 1 year period

Steps 3-5 for NUMBER of male and female buffalo

Step 3	Insert number of male/female buffalo in box N8 & O8
Step 4	Insert percent forage allocated to buffalo in box Q12
Step 5	Insert growing season favorability in box P16

Gender	Male Buffalo	Female Buffalo	Total Buffalo	Total Animal Units
Number of Bison	8	45	53	57

Water Requirement (gal/day)	
Low (15 gal)	795
Average (20 gal)	1060
High (30 gal)	1590

Higher percentage= Higher risk

Enter the percentage of forage allocated to the buffalo (25% recommended)	25%

If the growing season is unfavorable, enter 1 If the growing season is normal, enter 2 If the growing season is favorable, enter 3	1

Forage Availability	Forage Requirement	How Long The Buffalo Can Feed
482,991 pounds per year	624,150 pounds per year	0.77 Years
	52,155 pounds per month	9.3 Months
	1,710 pounds per day	282 Days

For 1 year, is there enough forage available to support

53 Buffalo?

NO	**Additional Food Required** **141,159 Pounds**

Figure 5. Right side of worksheet 4 of the Forage and Water Estimator Tool customized for Taos Pueblo.

The left side of worksheet 4 (Figure 4) shows that 20% forage allocation during an unfavorable year would support 35 AUY. This is equivalent to supporting 31 buffalo (4 male and 27 female) for 1 year. The right side of worksheet 4 (Figure 5) shows that if managers choose to have 8 male and 45 female buffalo and allocate 25% of forage production during an unfavorable year, then 141,159 pounds of additional forage would need to be supplied. The estimated water requirement is also shown for the 53 buffalo and ranges from 795 to 1,590 gallons of water per day.

Water Development Options

Current Water Options

Although there may be some risk to sustain the desired herd population during unfavorable years with respect to forage, without water there is no way to sustain the herd. Currently, there are two existing watering options within the pasture. The first option is an abandoned, non-working well. This well, East Well #2, has been inoperable since the early 1990s (InterTribal Buffalo Council, 2014). Rehabilitation of abandoned wells is not desirable due to high costs and uncertainty of water production. The dotted circles in Figure 6 represents a 1-mile radius around the abandoned wells and indicates that the current location is not optimized for access to the most productive soils.

This well is in the eastern portion of the pasture, which is a lower producing area based on soil productivity ratings. Because the well is in a low forage production area, overgrazing may occur during periods of high temperature and lead to even lower forage production in years to come.

The second, current option, is to develop surface water; there are several earthen dams located within the pasture. These earthen dams impound surface water; however, there are no continuous or abundant sources of flow through the pasture. Even if there was a reliable source to fill these dams, it was advised to not use them for several reasons. The annual evaporation rate in Taos County, New Mexico has been estimated to range from 43 to 56 inches, with 38 to 54 inches occurring between May and October (Garrabrant 1993). Extensive water loss of the impounded water due to evaporation and infiltration into the ground would consume whatever resources are available and would not favor sustainable practice. Other reasons include expensive cost to bury pipe below frost depth, potential algal blooms that can kill the buffalo, stagnant water, increased amounts of dissolved constituents that could negatively impact water quality, and the water becoming too hot.

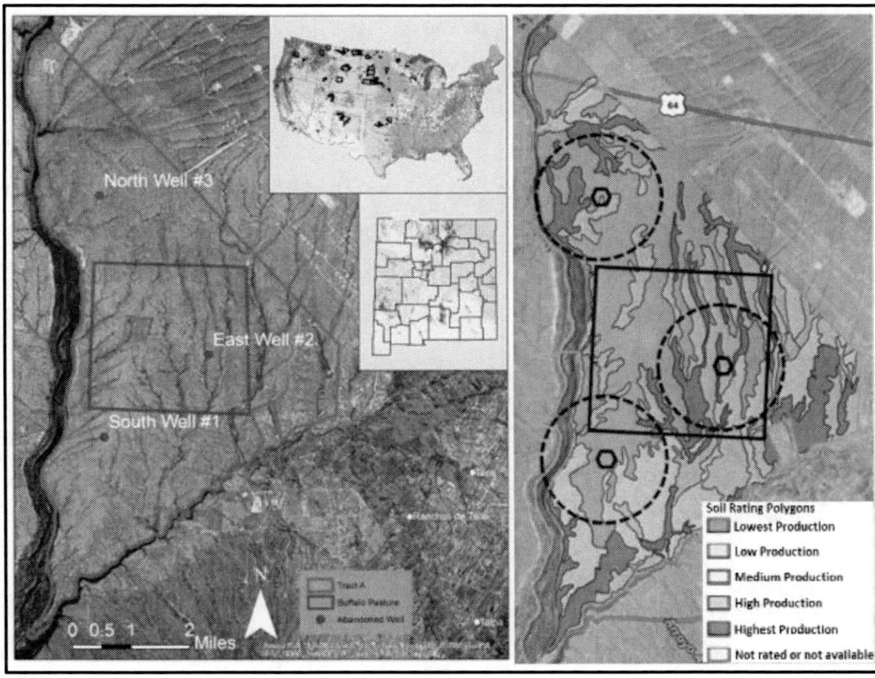

Figure 6. Abandoned wells in Taos Pueblo Tract A (left) and Web Soil Survey forage productivity (right).

Proposed Development Options

The proposed watering options included various locations for a single well and multiple wells. A new well was recommended over rehabilitating the abandoned wells as a new well allows the Tribe to place it in an optimum location with higher forage production or in an area that maximizes pasture usage. Refer to Wilder (2018) for details on the proposed development options. The option that was recommended was selected because its location maximizes the effective area for the higher forage producing areas during the hottest parts of the year, assuming a 1-mile travel distance by the buffalo. Because the recommended option is in the best forage production area, there is less chance of overgrazing, although it introduces the risk of soil compaction and forage degradation from repeated travel by buffalo to reach the water source, which may lead to lower forage production in future years.

The maximum water demand for 70 buffalo is estimated to be 2,100 gallons per day, based on 30 gallons per day per animal on days exceeding 90°F. Since there is no electricity in the pasture, a solar powered system is required for the well pump. To determine the necessary pumping capacity for the system, an assumed maximum demand in the summer over the shortest possible daylight period in the winter; since these two parameters never occur together, the resulting estimate should be more than adequate to meet current and future needs. Per the United States Naval Observatory Astronomical Applications Department, the day with the least daylight hours in Taos is December 21st, with an estimated time of 9.6 hours. This is conservative as the buffalo would not require as much water in December as they would in June. Therefore, to provide 2,100 gallons of water during a 9.6-hour window of daylight, a continuous flow rate of 3.65 gallons per minute (gpm) is required. This was determined to be feasible. The possibility of piping water in from a location near the pasture was also considered, but due to the large price was deemed not feasible and was dropped from further consideration. Refer to Wilder (2018) for several combinations of cost estimates for water development options.

Recommendations

1. A storage tank with 5,000-7,500 gallons capacity would provide sufficient reserves in case the watering system requires repair.
2. During unfavorable growing seasons, around 50 buffalo will be able to feed at 25% forage utilization. Although having more is possible, the risk of overgrazing starts to increase. If more than 50 are on the pasture, supplemental feed can be supplied to reduce pasture grazing.

Ponca Tribe of Nebraska

Introduction

The Ponca Tribe of Nebraska manages around 60 buffalo on 600 acres in Knox County in northeastern Nebraska. They manage buffalo on two pastures (Figure 7) one of which has surface water available, while the other has available surface water and gets water from a rural water supply. Ponca Tribe buffalo managers are concerned about water quality and associated detrimental impacts. Managing mineral excessiveness and deficiency is critical to Ponca buffalo management and water treatment may be necessary.

Figure 7. Map of Ponca's buffalo pastures.

This project focuses on the Agency Pasture (top of Figure 7), which has surface water from an intermittent stream with two ponds. The pasture also has a third pond in a low point away from the stream. Water quality samples were analyzed for this pasture following a buffalo necropsy report that indicated mineral imbalances in the buffalo.

A water quality analysis at 3 points along the stream, one of the ponds, and a wetland revealed high levels of calcium, sulfate, and manganese in the Agency Pasture water supply. Calcium ranged from 211 to 375 parts per million (ppm) in the stream, which is above the recommended range for livestock of 50-200 ppm. Sulfates ranged from 684 to 1,290 ppm in the stream and wetland, well above the recommended amount of less than 250 ppm. Manganese levels ranged from 0.097 to 1.065 ppm in all 5 water samples, well above the recommended limit of 0.05 ppm. Excess calcium, manganese, and sulfates can have a laxative effect on both buffalo and humans and may have contributed to copper deficiencies. For comparison, the West Knox Rural Water District water supply that the East Pasture uses met recommended quality standards as of 1991.

One of the buffalo that died had low levels of copper (1.4 ppm, normal range is 25-100 ppm), high levels of selenium (1.1 ppm, normal range is 0.25-0.5 ppm) and zinc (117 ppm, normal range is 25-100 ppm), and levels of molybdenum (0.8 ppm, no recommended range) that would interfere with copper absorption. The high selenium and zinc levels were likely related to the mineral supplement that had been used, and a new mineral supplement has been developed.

Interaction between the sulfates in the water supply and molybdenum may be contributing to copper deficiency. Unfortunately, it is currently unknown where the molybdenum is coming from. Water quality testing, and soil testing did not include results for molybdenum.

Forage and Water Supply and Demand

Total annual forage production, animal unit time, herd size, and water demand were estimated with the Forage and Water Estimator Tool based on spatial data from Ponca and the NRCS Web Soil Survey. Forage productivity by soil type from the Web Soil Survey was copied and pasted into the first worksheet (Figure 8).

Forage production is calculated from the soil productivity estimates. The total pasture forage production is calculated for each growing condition (Figure 9).

Annual forage production ranges from 376,254 – 524,625 pounds for the Agency Pasture, and from 961,135 – 1,553,508 pounds for the East Pasture, depending on the favorability of the growing season.

These pastures are estimated to support 3-24 AUY, and 9-71 AUY respectively, depending on growing season favorability and the percentage of annual forage production allocated to buffalo. The range of forage allocations presented represents a range of management choices consistent with generally accepted grazing management practices.

Worksheet 1- This worksheet is where the soil productivity data is entered

Web Soil Survey (NRCS) Range Soil Production Estimates for favorable, normal, and unfavorable growing seasons

Favorable Productivity					*Normal Productivity*					*Unfavorable Productivity*				
Summary by Map Unit — #1, Knox County, Nebraska (NE107)					**Summary by Map Unit — #1, Knox County, Nebraska (NE107)**					**Summary by Map Unit — #1, Knox County, Nebraska (NE107)**				
Map unit symbol	**Map unit name**	**Rating (pounds per acre per year)**	**Acres in AOI**	**Percent of AOI**	**Map unit symbol**	**Map unit name**	**Rating (pounds per acre per year)**	**Acres in AOI**	**Percent of AOI**	**Map unit symbol**	**Map unit name**	**Rating (pounds per acre per year)**	**Acres in AOI**	**Percent of AOI**
6606	Bazile loam, 6 to 11 percent slopes	3730	1	0.20%	6606	Bazile loam, 6 to 11 percent slopes	3234	1	0.20%	6606	Bazile loam, 6 to 11 percent slopes	2738	1	0.20%
6681	Crofton silt loam, 17 to 30 percent slopes, eroded	3348	36.4	6.10%	6681	Crofton silt loam, 17 to 30 percent slopes, eroded	2820	36.4	6.10%	6681	Crofton silt loam, 17 to 30 percent slopes, eroded	2038	36.4	6.10%

Figure 8. Forage and Water Estimator Tool customized for Ponca, worksheet 1.

Worksheet 2- This worksheet is where the total pasture productivity is calculated

This table gets the Web Soil Survey data for favorable normal and unfavorable conditions into one place and calculates the total pasture production. The productivity for each soil type is multiplied by the area in acres to get a production value for each soil type. The production values for all soil types in the pasture are then summed to get total pasture forage production.

Pasture number	*Map unit symbol*	*Map unit name*	*Forage Production (pounds)*		
			Favorable	*Normal*	*Unfavorable*
1	6606	Bazile loam, 6 to 11 percent slopes	3730	3234	2738
	6681	Crofton silt loam, 17 to 30 percent slopes, eroded	121867	102648	74183
	6693	Crofton-Nora complex, 2 to 6 percent slopes, eroded	16803	13950	9917
	6694	Crofton-Nora complex, 6 to 11 percent slopes, eroded	72512	60420	43157
	6753	Nora silt loam, 2 to 6 percent slopes	105332	85841	59675
	6767	Nora silty clay loam, 6 to 11 percent slopes	54137	44160	30746
	6789	Crofton-Nora complex, 11 to 17 percent slopes, eroded	81473	67687	48319
	8938	Simeon-Thurman complex, 6 to 30 percent slopes	31059	28356	24633
		Total Pasture Forage Production (pounds)	*486913*	*406295*	*293368*
2	3232	Lynch-Bristow complex, 11 to 30 percent slopes	82430	73220	65391
	3259	Meadin-O'Neill complex, 2 to 30 percent slopes	14950	13650	12090
	3287	Paka loam, 6 to 11 percent slopes, eroded	12400	11160	10230
	6681	Crofton silt loam, 17 to 30 percent slopes, eroded	416156	350526	253323
	6693	Crofton-Nora complex, 2 to 6 percent slopes, eroded	3933	3265	2321
	6694	Crofton-Nora complex, 6 to 11 percent slopes, eroded	299200	249305	178075
	6727	Thurman fine sandy loam, 2 to 11 percent slopes	32900	31020	28200
	6753	Nora silt loam, 2 to 6 percent slopes	194883	158821	110409
	8938	Simeon-Thurman complex, 6 to 30 percent slopes	9744	8896	7728
		Total Pasture Forage Production (pounds)	*1066595*	*899862*	*667767*

East Pasture

Figure 9. Forage and Water Estimator Tool customized for Ponca, worksheet 2.

Worksheet 3- This worksheet is where the pasture stocking capacity is estimated interms of Animal Unit Time

This set of tables estimate animal unit time that can be supported by the expected annual forage production estimate. I include a range of amounts of forage that could be allocated to bison with an upper limit of allocating half of the annual total pasture forage production to bison; this level of forage consumption is a generally accepted maximum. Allowed annimal unit time is given in animal unit days AUD, animal unit months AUM, and animal unit years AUY.

		Low Risk		Moderate Risk		High Risk
		Animal Unit Days (AUD)				
Growing Season Condition	*Total Pasture Forage Production (pounds)*	*10% of forage allocated to Buffalo*	*20% of forage allocated to buffalo*	*30% of forage allocated to buffalo*	*40% of forage allocated to buffalo*	*50% of forage allocated to buffalo*
Unfavorable	1345135	4484	8968	13451	17935	22419
Normal	1767685	5892	11785	17677	23569	29461
Favorable	2087640	6959	13918	20876	27835	34794

		Animal Unit Months (AUM)				
Growing Season Condition	*Total Pasture Forage Production (pounds)*	*10% of forage allocated to buffalo*	*20% of forage allocated to buffalo*	*30% of forage allocated to buffalo*	*40% of forage allocated to buffalo*	*50% of forage allocated to buffalo*
Unfavorable	1345135	147	294	441	588	735
Normal	1767685	193	386	580	773	966
Favorable	2087640	228	456	684	913	1141

This is for both pastures combined

		Animal Unit Years (AUY)				
Growing Season Condition	*Total Pasture Forage Production (pounds)*	*10% of forage allocated to buffalo*	*20% of forage allocated to buffalo*	*30% of forage allocated to buffalo*	*40% of forage allocated to buffalo*	*50% of forage allocated to buffalo*
Unfavorable	1345135	12	25	37	49	61
Normal	1767685	16	32	48	65	81
Favorable	2087640	19	38	57	76	95

Figure 10. Forage and Water Estimator Tool customized for Ponca, worksheet 3.

Risk of overgrazing is low when buffalo are allocated 10-25% of the total annual forage production, moderate at 25-40% of the total annual forage production, and high above 40% of the total annual forage production. When buffalo have consumed the allocated amount of forage, they should be fed hay or moved.

The number of buffalo that the pastures could support on an annual basis across a range of forage allocations were estimated for a herd of 15% males and 85% females. Figure 11 shows AUY and the number of buffalo for a range of forage allocations calculated by the Forage and Water Estimator tool and Figure 12 shows the estimated of the amount of additional forage necessary for 61 buffalo.

The left side of worksheet 4 (Figure 11) shows that 20% forage allocation during an unfavorable year would support 25 AUY. This is equivalent to supporting 22 buffalo (3 male and 19 female) for 1 year. The right side of worksheet 4 (Figure 12) shows that if managers choose to have 4 male buffalo and 22 female buffalo on the Agency Pasture and allocate 25% of forage production during an unfavorable year, there is only enough forage to support the buffalo for 135 days.

Additionally, if managers choose to have 5 male buffalo and 30 female buffalo on the East Pasture and allocate 25% of forage production during an unfavorable year, then there is only enough forage to support the buffalo for 290 days. Considering the buffalo are kept year-round in that herd structure, the Forage and Water Estimator Tool indicates that additional forage is being supplied or the pasture is being over-grazed. The estimated water requirement is also shown for the buffalo and ranges from 390 to 780 gallons of water per day for the Agency Pasture, and 525 to 1050 gallons of water per day for the East Pasture.

This worksheet represents the amount of buffalo over a 1 year period

Steps 1&2 for PERCENTAGE RATIO of male and female buffalo for BOTH pastures combined

If the exact amount of desired male and female buffalo are known, use Steps 3 and 4

Step 1	Insert percentage of male buffalo desired in box B8
Step 2	Insert percentage of female buffalo desired in box C8

Gender	Male Buffalo	Female Buffalo	Total
Percentage of Buffalo	15%	85%	100%

The following three tables provide the amount of male/female Buffalo that can be supported based on available forage and herd structure. It also shows the water requirement.

Low Risk | Moderate Risk | High Risk

Unfavorable Growing Season

% Forage Allocated	10%	20%	30%	40%	50%
Maximum Animal Units	12	25	37	49	61
Number of Buffalo	1 male 9 female	3 male 19 female	5 male 29 female	6 male 38 female	8 male 48 female
Water Requirement (gal/day)					
Low (15 gal)	150	330	510	660	840
Average (20 gal)	200	440	680	880	1120
High (30 gal)	300	660	1020	1320	1680

Normal Growing Season

% Forage Allocated	10%	20%	30%	40%	50%
Maximum Animal Units	16	32	48	65	81
Number of Buffalo	2 male 12 female	4 male 25 female	6 male 38 female	9 male 51 female	11 male 63 female
Water Requirement (gal/day)					
Low (15 gal)	210	435	660	900	1110
Average (20 gal)	280	580	880	1200	1480
High (30 gal)	420	870	1320	1800	2220

Favorable Growing Season

% Forage Allocated	10%	20%	30%	40%	50%
Maximum Animal Units	19	38	57	76	95
Number of Buffalo	2 male 15 female	5 male 30 female	7 male 45 female	10 male 60 female	13 male 75 female
Water Requirement (gal/day)					
Low (15 gal)	255	525	780	1050	1320
Average (20 gal)	340	700	1040	1400	1760
High (30 gal)	510	1050	1560	2100	2640

Figure 11. Left side of worksheet 4 of the Forage and Water Estimator Tool customized for Ponca.

This worksheet represents the amount of buffalo over a 1 year period

Steps 3-5 for NUMBER of male and female buffalo for each pasture SEPERATELY

Step 3	Insert number of male/female buffalo in box N8 &O8 and N9 & O9
Step 4	Insert percent forage allocated to buffalo in box N20 & Q20
Step 5	Insert growing season favorability in box P24

Pasture	Male Buffalo	Female Buffalo	Total Buffalo	Total Animal Units
Agency Pasture	4	22	26	28
East Pasture	5	30	35	37.5

Agency Pasture

Water Requirement (gal/day)	
Low (15 gal)	390
Average (20 gal)	520
High (30 gal)	780

East Pasture

Water Requirement (gal/day)	
Low (15 gal)	525
Average (20 gal	700
High (30 gal)	1050

Higher percentage= Higher risk

Enter the percentage of forage allocated to the buffalo (25% recommended)

Agency Pasture	25%	East Pasture	25%

If the growing season is unfavorable, enter 1
If the growing season is normal, enter 2
If the growing season is favorable, enter 3

2

Agency Pasture

Forage Availability	Forage Requirement	How Long The Buffalo Can Feed
113,204 pounds per year	306,600 pounds per year	0.4 Years
	25,620 pounds per month	4.4 Months
	0,840 pounds per day	134.8 Days

East Pasture

Forage Availability	Forage Requirement	How Long The Buffalo Can Feed
326,539 pounds per year	410,625 pounds per year	0.8 Years
	34,313 pounds per month	9.5 Months
	1,125 pounds per day	290.3 Days

For 1 year, is there enough forage available to support

Agency Pasture 26 Buffalo?	East Pasture 35 Buffalo?
NO	**NO**
Additional Food Required	Additional Food Required
193,396 Pounds	84,086 Pounds

Figure 12. Right side of worksheet 4 of the Forage and Water Estimator Tool customized for Ponca.

Water Quality Response Options

Surface water in the Agency Pasture has high levels of calcium, manganese, and sulfates. Water treatment techniques that can remove or reduce the amount of these dissolved solids include reverse osmosis (RO), ion exchange (IE), and distillation (Minnesota Department of Health and Well Management Section Environmental Health Division 2016). These water treatments can create nearly 100% pure water that can then be mixed with the untreated water at a defined ratio that meets water quality guidelines and minimizes the amount of treated water required. Other water treatment options such as water softeners, carbon filtration, or sediment filtration do not remove sulfate. All three of the water treatment systems that can remove sulfates are expensive in terms of capital costs, operating costs, and maintenance costs.

Reverse Osmosis is a water treatment technology that forces water through a semipermeable membrane, which allows water molecules to pass through it while blocking molecules such as sodium. Reverse osmosis can remove 95-99% of the dissolved salts, however, it requires high pressure, or energy. Additionally, it requires significantly more source water than what is treated. For instance, to produce 1,000 gallons of clean water, 2000-4,000 gallons of untreated water is required, depending on the treatment system configuration. Less water can be used, but then the cost of treatment goes up, as it requires more energy. These systems are expensive to purchase and have daily usage costs as well as yearly maintenance costs. The RO membranes can be fouled by bacteria, particles such as sand or clay, natural organic material, chlorine, or mineral precipitants such as those formed by calcium, magnesium, and sulfates. Typically, reverse osmosis systems require multiple pretreatments such as filtration, scale inhibitor addition, and pH adjustment. Based on the complexity of the treatment system with pretreatment requirements, it is not recommended for surface water sources.

Ion Exchange is a water treatment technology that removes unwanted contaminants by replacing them with different ions. Essentially, undesirable ions such as magnesium, are replaced with ions such as sodium or potassium, which may be more desirable. Ion exchange resins start with a high concentration of the replacement ions, which are released into the water when the undesirable ions bind to the resin. Ion exchange resins can be regenerated by washing the system with a high concentration of the replacement ions. This system is like a water softening system in that ion exchange systems require purchase of regenerative material, which can be expensive. Additionally, like reverse osmosis, this method often requires pretreatment and more water than what is treated.

Distillation treats water by evaporation and condensation. Boiling water produces steam that does not have impurities. This steam is then captured and condensed to produce pure water. Distillation is an energy intensive process and will require electricity. Assuming a yearly average water temperature of 70°F, it would take approximately $67 in electricity per day to bring 2,000 gallons of water to boiling temperature. The annual estimated expense of $24,000 per year does not consider additional electricity needed to keep the water boiling or to condense the water vapor, which typically is done by a heat exchanger/refrigeration unit. This option does not seem feasible and therefore is not further considered in this study.

Water Quality Analysis

Reverse Osmosis

The following tables show the original water quality, treated water quality assuming a 95% removal efficiency, and blended water qualities with various blending ratios. Green represents good water quality, yellow represents borderline water quality, and red represents unacceptable water quality (level exceeds limits). These values were assigned based on EPA recommended contaminant levels, as shown in Table 1.

From Table 2, it is apparent that water quality in the Ponca Tribe Agency Pasture poses challenges for supporting buffalo. Concentrations for six of the ten constituents exceed recommended levels. Manganese (Mn) is upwards of 21 times higher and sulfate (SO_4) is upwards of 5 times higher. Calcium in the Stream_in sample is almost twice the recommended level. One of the proposed treatment options was reverse osmosis, which has a removal efficiency of 95-99%. To be conservative, a removal efficiency of 95% was chosen. Assuming the necessary pretreatment steps were done to allow reverse osmosis to operate correctly, Table 3 shows the estimated treated water quality. It is important to note that this is just an estimate assuming 95% reduction in all constituents. Actual operation may result in higher removal of some constituents and lower removal of others.

Table 1. Water quality recommended concentration levels for different parameters tested in the Ponca water quality analysis

	Na (ppm)	Ca (ppm)	Mg (ppm)	NO3-N (ppm)	SO4 (ppm)	TDS (ppm)	Fe (ppm)	Mn (ppm)	Cl (ppm)	F (ppm)
EPA- Human	< 100		< 80	< 10	< 250		< 0.3	< 0.05	< 250	
Livestock	< 100	50-200	< 80	< 25	< 250	100-500 good, > 3000 bad	< 0.3	< 0.05	< 250	< 2
MidwestLab caution level	< 100	< 80	< 30	< 10	< 400	< 500	< 0.3	< 0.05	< 200	< 4

Table 2. Original water quality data provided by Ponca tribe of Nebraska for 5 surface water samples collected in the Agency Pasture in 2017

	Na (ppm)	Ca (ppm)	Mg (ppm)	NO3-N (ppm)	SO4 (ppm)	TDS (ppm)	Fe (ppm)	Mn (ppm)	Cl (ppm)	F (ppm)
Wetland	33.8	188	43.9	n.d.	684	936	0.18	0.511	22	0.2
Pond 4	17.4	79.4	10.7	n.d.	165	376	0.13	0.492	11	0.1
Stream_in	64.5	375	90.4	n.d.	1290	1640	0.39	1.065	47	0.5
Stream_mid	37.4	211	51.4	0.3	708	962	0.05	0.179	22	0.2
Stream_out	42.1	237	56.7	n.d.	782	1040	0.06	0.097	25	0.3

Table 3. Potential water quality for the Agency Pasture after water treatment, assuming 95% removal efficiency

	Na (ppm)	Ca (ppm)	Mg (ppm)	NO3-N (ppm)	SO4 (ppm)	TDS (ppm)	Fe (ppm)	Mn (ppm)	Cl (ppm)	F (ppm)
Wetland	1.7	9	2.2	n.d.	34	47	0.009	0.027	1.1	0.01
Pond 4	0.9	4	0.5	n.d.	8	19	0.007	0.025	0.55	0.005
Stream_in	3.2	19	4.5	n.d.	65	82	0.020	0.053	2.35	0.025
Stream_mid	1.9	11	2.6	0.015	35	48	0.003	0.009	1.1	0.01
Stream_out	2.1	12	2.8	n.d.	39	52	0.003	0.0049	1.25	0.015

The water quality following reverse osmosis is high, as the process removes almost all dissolved solids. This estimate shows that the constituent concentration for each water source, except for manganese in the Stream-in, does not exceed recommended levels. This water may not have high enough concentration of certain minerals, such as calcium, for the buffalo. One way to restore the mineral concentration without exceeding levels is to blend some of the untreated water (raw water) with the treated water. This method has two benefits: 1) it increases the mineral concentration to provide important nutrients and 2) it reduces the amount of water that needs to be treated. Instead of treating 1,000 gallons every day, 750 gallons may be treated and blended with 250 gallons of raw water. This blending ratio depends on the treated water quality and the raw water quality. Once combined, the blended water should not exceed any of the concentration limits.

By summing the treated water volume multiplied by treated water concentration and the raw water volume multiplied by the raw water concentration, and dividing by the total volume, the blended concentration is determined. An example of this is shown below for sodium (Na). It should be noted that 1 ppm is equivalent to 1 mg/L.

Equation 1: Average Concentration of Blended Water

$$Na\ Concentration\ (ppm) = \left(\frac{(Treated\ water\ (gal) * Treated\ water\ Na\ (ppm)) + (Raw\ water\ (gal) * Raw\ water\ Na\ (ppm))}{Treated\ water\ (gal) + Raw\ water\ (gal)}\right)$$

This was done for each constituent in the water quality analysis for 3 different blending ratios; 4:1, 3:1, and 2:1 (gallons treated water: gallon raw water). Table 4 shows the estimated water quality resulting from a 4:1 blend.

Except for manganese, all constituents do not exceed recommended levels. To produce 1,000 gallons of water a day, approximately 800 gallons of treated water from Stream_mid or Stream_out should be blended with 200 gallons of water from the respective source. Table 5 shows the estimated water quality resulting from a 3:1 blend.

Table 4. Potential water quality for the Agency Pasture after water treatment, assuming 95% removal efficiency and blending water at a ratio of 4:1

	Na (ppm)	Ca (ppm)	Mg (ppm)	NO3-N (ppm)	SO4 (ppm)	TDS (ppm)	Fe (ppm)	Mn (ppm)	Cl (ppm)	F (ppm)
Wetland	8.1	45	10.5	n.d.	164	225	0.043	0.123	5.28	0.048
Pond 4	4.2	19	2.6	n.d.	40	90	0.031	0.118	2.64	0.024
Stream_in	15.5	90	21.7	n.d.	310	394	0.094	0.256	11.28	0.12
Stream_mid	9.0	51	12.3	0.072	170	231	0.012	0.043	5.28	0.048
Stream_out	10.1	57	13.6	n.d.	188	250	0.014	0.023	6	0.072

Table 5. Potential water quality for the Agency Pasture after water treatment, assuming 95% removal efficiency and blending water at a ratio of 3:1

	Na (ppm)	Ca (ppm)	Mg (ppm)	NO3-N (ppm)	SO4 (ppm)	TDS (ppm)	Fe (ppm)	Mn (ppm)	Cl (ppm)	F (ppm)
Wetland	9.7	54	12.6	n.d.	197	269	0.052	0.147	6.33	0.06
Pond 4	5.0	23	3.1	n.d.	47	108	0.037	0.141	3.16	0.03
Stream_in	18.5	108	26.0	n.d.	371	472	0.112	0.306	13.51	0.14
Stream_mid	10.8	61	14.8	0.086	204	277	0.014	0.051	6.33	0.06
Stream_out	12.1	68	16.3	n.d.	225	299	0.017	0.028	7.19	0.09

Table 6. Potential water quality for the Agency Pasture after water treatment, assuming 95% removal efficiency and blending water at a ratio of 2:1

	Na (ppm)	Ca (ppm)	Mg (ppm)	NO3-N (ppm)	SO4 (ppm)	TDS (ppm)	Fe (ppm)	Mn (ppm)	Cl (ppm)	F (ppm)
Wetland	12.4	69	16.1	n.d.	251	343	0.066	0.187	8.07	0.07
Pond 4	6.4	29	3.9	n.d.	61	138	0.048	0.180	4.03	0.04
Stream_in	23.7	138	33.1	n.d.	473	601	0.143	0.391	17.23	0.18
Stream_mid	13.7	77	18.8	0.11	260	353	0.018	0.066	8.07	0.07
Stream_out	15.4	87	20.8	n.d.	287	381	0.022	0.036	9.17	0.11

Again, except for manganese, all constituents do not exceed recommended levels. To produce 1,000 gallons of water a day, approximately 750 gallons of treated water from Stream_out would have to be blended with 250 gallons of raw water from Stream_out. Table 6 shows the estimated water quality resulting from a 2:1 blend.

At a 2:1 blend, the manganese concentration resulting from blending treated water from Stream_out does not exceed recommended levels, but sulfate is becoming a potential issue. Depending on the source of the recommendation, a sulfate concentration greater than 250 ppm or greater than 400-ppm are considered problematic. If based on the 250-ppm recommendation, a 2:1 blend results in a sulfate concentration that just barely exceeds. If based on the 400-ppm recommendation, the sulfate concentration is fine. To produce 1,000 gallons of water a day, approximately 667 gallons of treated water from Stream_out would have to be blended with 333 gallons of untreated water from Stream_out. The raw water quality is not fixed but can vary with the seasons, annual precipitation, temperatures, and other factors. Regular testing of the raw water sources is needed to properly balance the mixing ratios and ensure quality water.

4.2.4.2. Ion exchange

Ion Exchange is different than reverse osmosis in that it replaces ions rather than removing them. EvoQUA Water Technologies, one of several companies offering ion exchange products, suggests that ion exchange treatment of water that has iron or manganese has limited application due to the tendency of the ions to react with oxygen. This reaction can cause plugging of the resin, buildup of precipitants on the resin, and sometimes irreversible fouling. However, there is a way to reduce the iron and manganese concentration before it goes into an ion exchange system by pretreating the water through oxidation. Oxidation involves aeration or introducing an oxidizing agent, such as chlorine, which will react with iron or manganese and form an insoluble particle, which can then be physically separated through filtration. Even so, several issues must be taken into consideration. Aeration, or blowing or spraying air into the water, will increase the treatment cost, including initial costs for the equipment, electricity costs to power the blower or compressor, and maintenance costs. Introducing an oxidizing agent such as chlorine adds the cost of chemicals and can potentially generate harmful byproducts. Chlorine reacts with natural organic material in the water produces disinfection byproducts; some are carcinogenic to humans. Their effects on buffalo are unknown.

4.2.5. Recommendations

Water quality in the Agency Pasture is a complex issue and should be fully analyzed before making any water development decisions. The water treatment systems that can treat the water to acceptable quality for the buffalo are expensive, complex, and require a lot of maintenance. Reverse osmosis is an innovative technology and works very well; however, it is sophisticated and requires a lot of knowledge and time to ensure proper operation.

Based on the current water quality data, it is believed that multiple pretreatment steps would be required as well as continual water quality monitoring. Ion exchange is also an advanced technology and is widely used in industry. Again, this technique requires a lot of knowledge and time to ensure proper operation. Additionally, due to the "swapping" of ions, which is how the system works, it is not clear if the high concentration of the replaced ions will affect the buffalo. As this water quality issue is so complex and a full analysis was not completed, it is unwise to make a water treatment recommendation. However, other recommendations for the Agency Pasture can be seen below.

1. A full water analysis and treatment option analysis needs to be done.
2. The water and soil should be tested for molybdenum. This element was identified in a buffalo necropsy report; its presence with sulfates can contribute to copper deficiencies. It is important to identify the source of molybdenum to manage copper absorption.
3. If the buffalo in the Agency Pasture continue to die, move them to the East Pasture and grow hay on the Agency Pasture. If this is done, less hay would have to be purchased. Instead of limiting the buffalo to 25% of the forage while grazing, 100% of the forage can be used

if hayed. This will offset some of the forage requirement in the now larger herd sized East Pasture and allow time to develop a plan for the Agency Pasture.

4. Since treating the water presents several challenges, another possibility is bringing a rural water supply to the Agency Pasture like the East Pasture. However, this is only feasible if there is a rural water main nearby as distances of several miles can be cost prohibitive. If rural water was supplied, the existing water sources on the Agency Pasture would need to be fenced off so that the buffalo cannot access it.

4.3. Model Builder Forage Estimator Automated Model

This section shows the use of the Automated Forage and Water Estimator Tool, as a comparison to how it was done manually in the pilot projects. Ease of use was an important objective when creating the automated model as it would not make sense to use if it was more complicated or time consuming than the manual tool.

Additionally, it was essential that the automated model produce results that were accurate, such as matching the results of the manual tool. For ease of comparison, Taos Pueblo was the example used again. The data being used by the model comes from the Web Soil Survey and should therefore match with the original tool.

4.3.1. Using the Automated Model to Estimate Forage and Water Supply and Demand

The Forage Estimator Model (Figure 13) performs a series of steps that leads to a calculated estimate of the total annual forage for favorable, normal, and unfavorable years.

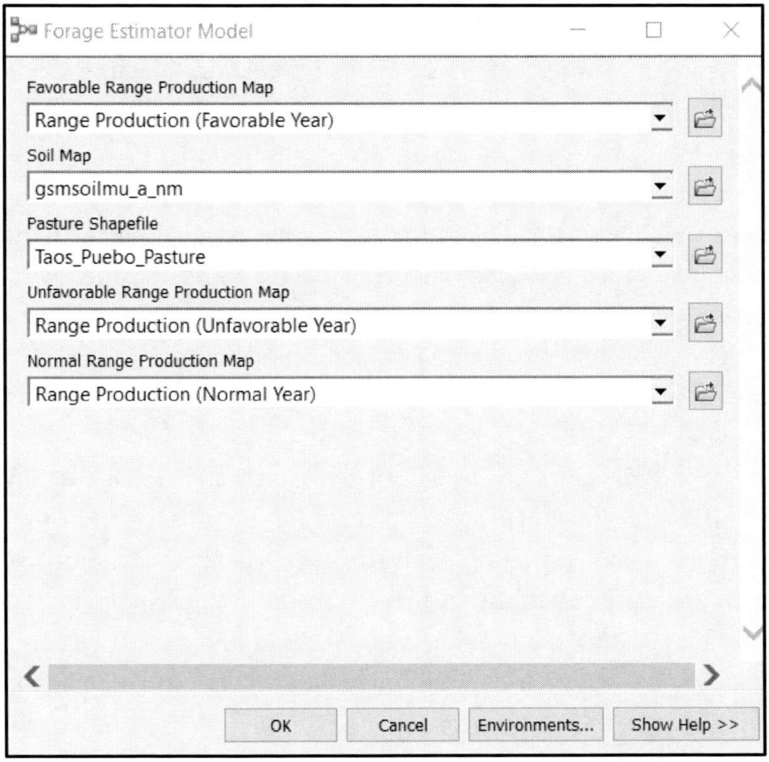

Figure 13. Screenshot of the ModelBuilder run interface for the Forage Estimator Model.

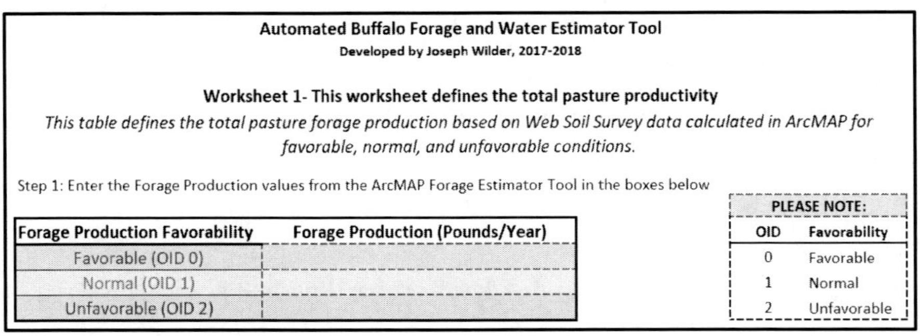

Automated Buffalo Forage and Water Estimator Tool

Developed by Joseph Wilder, 2017-2018

Worksheet 1- This worksheet defines the total pasture productivity

This table defines the total pasture forage production based on Web Soil Survey data calculated in ArcMAP for favorable, normal, and unfavorable conditions.

Step 1: Enter the Forage Production values from the ArcMAP Forage Estimator Tool in the boxes below

Forage Production Favorability	Forage Production (Pounds/Year)
Favorable (OID 0)	
Normal (OID 1)	
Unfavorable (OID 2)	

PLEASE NOTE:

OID	Favorability
0	Favorable
1	Normal
2	Unfavorable

Figure 14. Screenshot of the automated Forage and Water Estimator Tool, worksheet 1.

The model uses soil maps and range productivity maps to calculate how much forage can be produced for a given area. The Forage Estimator Model automates the process of uploading a pasture shapefile and the supporting files into Web Soil Survey, opening Soil Data Explorer, selecting Vegetative Productivity, selecting favorable, normal, or unfavorable conditions, viewing the ratings, copying/pasting the tables into excel, performing calculations to determine the total forage production per soil type, and performing calculations to determine the forage production for the entire pasture. This tool results in an output table that represents how many pounds of forage the pasture will produce in one year during favorable, normal, and unfavorable conditions. Production values from this tool are entered into worksheet 1 of the Automated Buffalo Forage and Water Estimator Tool (Figure 14).

The second and third worksheet of the Automated Forage and Water Estimator Tool are identical as worksheets three and four from the pilot projects. The automated version's total forage is calculated automatically in ArcMap based on the buffalo pasture. Essentially, the automated version simplifies the first two worksheets as seen in the pilot projects as it eliminates manual data retrieval and manual calculations of data from the Web Soil Survey's online application.

CONCLUSION

Estimates made by the Forage and Water Estimator Tool unique to each pasture and actual forage production may be worse than estimated as there are many areas of potential error, such as (1) degraded land condition, (2) altered plant communities, (3) inaccuracy of original soil mapping, and (4) overestimated productivity values.

The potential usefulness of these tools were illustrated by its application in the pilot projects. Tribal resource managers can input management choices into the tool to address their specific goals, which results in; (1) an estimate of how many buffalo the pasture can support, (2) how much additional forage may need to be provided, and (3) how much water will be required. These results are then used to inform water development choices, such as storage tank size, and herd structure, such as gender ratio. Water development and herd structure significantly affects climate change vulnerability assessments and drought adaptation. For instance, if the proper size storage is chosen which can meet peak demand for at least 3 consecutive days, exposure to drought will be low. If a herd size is chosen that represents around 50% of forage during a favorable year, then exposure to climate change will be high since the same herd structure during unfavorable years may be 100% utilization. Specifically, this tool demonstrated that the Taos Pueblo pasture should be able to maintain a herd of about 40 buffalo annually with low risk based on unfavorable conditions. The tool also shows that the pasture can support a larger herd, such as the 50-70 buffalo that they would like, but become vulnerable to climate change. For the Ponca Tribe, this tool demonstrated that their pastures should be able to maintain a combined herd of about 22 buffalo annually with low risk based on unfavorable conditions. This is substantially lower than the 60 buffalo they currently manage.

The tool also shows that to maintain their current herd of around 60 buffalo and keep forage allocation around 25%, they must supply around 280,000 pounds of additional forage. Supporting their current herd without supplying additional forage could be done, but the pasture would become vulnerable to climate change as total annual forage allocation would be around 55% during an unfavorable year. Lastly, this tool helped in understanding how much water their buffalo require, which influenced treatment considerations.

The ArcMap based Forage Estimator Model was not used in a pilot project setting as it was developed after the initial project, but the potential usefulness of it is significant. With just a few steps, these tools can estimate the total annual forage production for a given land segment for favorable, normal and unfavorable growing seasons. The usefulness of these tools goes beyond raising buffalo for Tribes and can be easily modified and used for cattle, sheep, goats, horses, and other livestock. Perhaps most important towards the objectives of this research, these tools allows Tribal resource managers to make management decisions and instantly see the numerical outcome of those choices, which should lead to more informed management practices.

ACKNOWLEDGMENTS

The authors would like to thank the Intertribal Buffalo Council for funding the original project from which this study evolved. Additional thanks would like to be given to following people from South Dakota School of Mines and Technology, whom contributed to the original project, *InterTribal Buffalo Council Climate Change Vulnerability Assessment and Effective Drought Management Adaptation* (2017).

- Dr. Bill Capehart
- Dr. Maribeth Price
- Mark Yeske
- Danielle Tourtillott.

REFERENCES

Alberta Agriculture and Forestry. 2009. *Farm Water Supply Requirements*. Agdex document 716. Available online at http://www1.agric.gov.ab.ca/$department/deptdocs.nsf/all/agdex1349#requirements.

Bedell, T. 1998. Glossary of terms used in range management. *Society for Range Management*, Denver, CO. 32pp.

Chowanski, K., Capehart, B., Price, M., Kenner, S., Gadhamshetty, V., Wilder, J., Yeske, M., and Tourtillott, D. (2017). *InterTribal Buffalo Council Climate Change Vulnerability Assessment and Effective Drought Management Adaptation.*

Duesterhaus, J., J. Ham, C. Owensby, and J. Murphy. 2008. Water balance of a stock-watering pond in the Flint Hills of Kansas. *Rangeland ecology & management* 61:329-338.

Garrabrant, L. A. 1993. *Water Resources of Taos County*, New Mexico. US Department of the Interior, US Geological Survey.

Holechek, J. L., 1988. An approach for setting the stocking rate. *Rangelands*, *10*(1), pp. 10-14.

Holechek, J. L. and Pieper, R. D., 1992. Estimation of stocking rate on New Mexico rangelands. *Journal of Soil and Water Conservation*, *47*(1), pp. 116-119.

Holechek, J. L., R. D. Pieper, and C. H. Herbel. 1989. *Range management: principles and practices.* Prentice-Hall, Englewood Cliffs, New Jersey.

InterTribal Buffalo Council. 2014. *Taos Pueblo 2014 Buffalo Expansion Feasibility Study*. Second Draft.

Kohl, M. T., P. R. Krausman, K. Kunkel, and D. M. Williams. 2013. Bison versus cattle: are they ecologically synonymous? *Rangeland ecology & management* 66:721-731.

McNeeley, S. M., T. A. Beeton, and D. S. Ojima. 2016. Drought risk and adaptation in the interior United States: understanding the importance of local context for resource management in times of drought. *Weather, Climate, and Society* 8:147-161.

Minnesota Department of Health, and Well Management Section Environmental Health Division. 2016. *Sulfate in Well Water*. Available http://www.health.state.mn.us/divs/eh/wells/waterquality/sulfate.html.

National Drought Mitigation Center. 2012. *Managing Drought Risk on the Ranch A Planning Guide for Great Plains Ranchers*. 39.

Nippert, J. B., T. S. Culbertson, G. L. Orozco, T. W. Ocheltree, and B. R. Helliker. 2013. Identifying the water sources consumed by bison: implications for large mammalian grazers worldwide. *Ecosphere* 4:1-13.

Polley, H. W., D. D. Briske, J. A. Morgan, K. Wolter, D. W. Bailey, and J. R. Brown. 2013. Climate change and North American rangelands: trends, projections, and implications. *Rangeland ecology & management* 66:493-511.

Soil Survey Staff, Natural Resources Conservation Service, and United States Department of Agriculture. *Web Soil Survey*. Available online at https://websoilsurvey.sc.egov.usda.gov/.

United States Department of Agriculture, and Natural Resources Conservation Service. 1997. *National Range and Pasture Handbook*. USDA, NRCS Washington, DC.

University of Nebraska. 2018. *Overgrazing and Drought*. Available online at http://drought.unl.edu/ranchplan/DroughtBasics/GrazingDrought/OvergrazingDrought.aspx.

Wilder, Joseph. 2018. *The Development of a Buffalo Forage and Water Estimator Tool That Can Be Used to Inform Management Decisions and Assist Climate Change Vulnerability Assessments and Drought Management Adaptation*. MS thesis, South Dakota School of Mines and Technology.

Winchester, C., and M. Morris. 1956. Water intake rates of cattle. *Journal of Animal Science* 15:722-740.

In: Sustainable Water
Editors: V. Gnaneswar Gude et al.
ISBN: 978-1-53617-335-2

Chapter 7

RESOURCE RECOVERY FROM REVERSE OSMOSIS CONCENTRATE AS A SOLUTION TO WATER CRISIS: A TECHNOLOGICAL ASSESSMENT

***Tabish Nawaz*[1,4] *and Thinesh Selvaratnam*[1,2,3,*]**

[1]Center for Advances in Water & Air Quality
Lamar University, Beaumont, TX, US
[2]Center for Midstream Management and Science
Lamar University, Beaumont, TX, US
[3]Department of Civil & Environmental Engineering
Lamar University, Beaumont, TX, US
[4]Environmental Science and Engineering Department,
Indian Institute of Technology Bombay, Powai,
Mumbai, Maharashtra, India

---

[*] Corresponding Author's Email: tselvaratnam@lamar.edu.

ABSTRACT

The rising world population, increasing urbanization, industrialization, and climate change have led to a global water crisis. Reuse of wastewater is essential to address the ongoing water crisis. However, water reuse requires technological, financial and social support. The emerging paradigm of resource recovery is a potential solution to make water reuse feasible and sustainable. Reverse osmosis (RO) concentrate is a waste stream discharged by RO desalination process plants. RO concentrate streams are rich in minerals and salts, which when economically recovered can bring down the cost of desalination and water. RO desalination is also considered crucial for increasing potable water availability. RO concentrate discharge, therefore, is expected to increase, which implies an increased risk of environmental harms posed by its release into the environment. Moreover, minerals/salts recovery from RO concentrate would lead to potential economic gains. In this article, the case for resource recovery from RO concentrate is made. The key material species found in RO concentrate are identified for their recovery and economic potential. The available methods reported in the literature are reviewed for their suitability for resource recovery from RO concentrate. The reviewed methods include chemical precipitation, crystallization, evaporation, membrane-based methods, adsorption/ion-exchange and few miscellaneous methods such as algal methods, and eutectic freeze point crystallization. For recovering high ionic strength species from RO concentrate chemical precipitation, crystallization and membrane-based methods can be useful; trace minerals' recovery can be achieved via ion-exchange, adsorption, electrodialysis, and liquid-liquid extraction-based methods. The species that can be profitably recovered include Na, K, Ca, Mg, Rb, In, Cs, Ge, Li, Sr, Br, and U.

1. INTRODUCTION

1.1. Global Water Crisis and Resource Recovery

Increasing global population, industrialization and agricultural activities, pollution, urbanization, and climate change have impacted the water availability for a large fraction of the world's population. According to a report published by United Nations University – Institute for Water, Environment, and Health (UNU – INWEH) in 2017, up to 40% of the world population will be living in severely water-stressed regions by 2035 [1]. Additionally, the scarcity would be deeply experienced among the poorest countries, in rural regions, and within urban centers among low-income groups and unregularized settlements. The WHO – UNICEF Joint Management Program Report (JMP) also published in 2017 reports that 844 million people in 2015 lack even a basic drinking water facility. Also, 159 million people collected drinking water directly from surface water sources; 58% of these people lived in Sub-Saharan Africa. The use of water without adequate treatment leads to several diseases such as diarrhea, cholera, dysentery, typhoid, and polio [2]. The increasing depletion and further aggravation of the problem caused by the contamination of available water resources have led to a global water crisis. Since agricultural activities consume approximately 70% of the total water withdrawal, the water crisis, linked to climate change and increasing population, also closely affects global food security [3].

The rising population and technological advancement have led to an increase in the scale of agricultural, industrial and social/household activities. The result has been a generation of huge amounts of wastewater (316 km^3/year of municipal wastewater and 632 km^3/year of industrial wastewater worldwide) [4] containing considerable varieties and quantities of chemical compounds. These substances when economically extracted, can lead to both environmental and economic benefits. Additionally, the used water might contain pollutants and pathogens. Therefore, the treatment of wastewater before reuse, depending on the nature of reuse – potable or non-potable, is critical. The treatment would then further add to the cost of water. One potential solution to offsetting the water cost is by recovering the pollutants from the wastewater as a marketable commodity. Therefore, resource recovery when successfully implemented would not only remove pollutants from waste streams but also recover them as useful economic products, and thus bring down the cost of produced water.

Resource recovery practice has transformed the notion of wastewater from one of a waste to that of a resource. This paradigm shift has led to view wastewater as a renewable resource for recoverable constituents such as carbon, nutrients, heavy metals, nutrients in addition to energy and water [5].

Recovering resources from waste streams are challenging. The crucial steps include identification of the resources and their markets, documenting available technologies, and development of cost-effective treatment methodologies. Moreover, it is critical to identify future commodities and value-added final products (e.g., biopolymers, trace metals, chemicals, nutrients (apart from nitrogen and phosphorus)), which would lead the wastewater treatment sector into future [5]. In the past few years, the desalination of seawater, particularly one based on reverse osmosis (RO) membrane systems, has increasingly been used, which leads to large-scale discharge of its waste streams known as RO concentrate [6]. The RO concentrate is typically discharged back to oceans and may pose environmental issues in the long run [7]. Moreover, treating it as a resource and recovering useful products out of it can not only bring economic dividends but also augment water recovery and supply.

1.2. RO Desalination Concentrates – Water Reuse and Resource Recovery Potential

Desalination technologies have been widely used to augment water supplies, particularly in areas where freshwater supply is scarce. An increase in clean water demand and reduction in desalination cost have further increased the global desalination capacity [6]. Desalination processes broadly fall into two major categories – thermal processes and membrane processes. Thermal processes include multi-effect distillation and multi-stage flash distillation, whereas membrane processes are predominantly based on reverse osmosis technologies. While thermal desalination remains the primary technology of choice in the Middle East, the number of RO desalination facilities has shown an increasing trend worldwide. Presently, over 15,000 desalination plants are in operation globally, and roughly 50% of these plants chiefly employ RO technologies. In the United States, RO technologies are utilized by 69% of the desalination plants and contribute to

84% desalination capacity. RO technologies employ semi-permeable membranes that separate a solution into two streams – permeate (containing the purified water) and concentrate (the portion that consists of salts and retained compounds) [8].

Due to the lower cost of water production than thermal processes (typically one-half to one-third of the cost of distillation) [9], RO technologies are considered key to increasing freshwater supplies from seawater sources. A significant increase in desalination capacity in recent times indicate a new trend in seawater desalination plants with production greater than 100,000 m^3/day [8]. However, there are multiple drawbacks associated with capital and operating costs and RO concentrate management and disposal that limit wider applications of RO technologies in inland and rural communities. The characteristics of RO concentrate depend on the feed water quality, water recovery rate, pretreatment methods, and cleaning compounds used [10]. In seawater desalination plants, the water recovery rate is typically 50 – 55% [11], therefore, the constituents' concentration in the concentrate is found to be double or higher than that of feed water supply. In coastal desalination units, concentrates are directly discharged to seawater; the representative discharge volume for large desalination units is 330,000 m^3/day in Ashkelon, Israel, 136,000 m^3/day in Tuas Seawater Desalination Plant in Singapore and 64,000 m^3/day in Larnaka facility in Cyprus and majority of large desalination units in Spain, Australia, and the Middle East [12].

Studies indicate adverse environmental impacts of concentrate discharge on aquatic organisms and corals, pH fluctuation, salinity change affecting dissolved oxygen levels, and heavy metal contamination of oceans from discharge pipe corrosion [7]. Additionally, there is a financial cost associated with discharging RO concentrate, roughly accounting for 5 – 33% of total desalination cost [13]. Therefore, several methods such as disposal to a combined sewer, evaporation ponds, deep well injections have been put into practice [8]. However, all the disposal options represent a loss of water and incur additional plant costs. Apart from the individual negative aspects of each disposal method, each is lacking from the point of additional water recovery that is possible, since all the water in the concentrate is lost during discharge. Enhanced water recovery utilizing RO concentrate can increase the RO process efficiency and bring down the cost of freshwater production. The proper utilization of RO concentrate can not only increase the water recovery by RO methods but also mitigate a potential environmental hazard caused due to its direct disposal in the water bodies. Studies report that the ultimate achievement in concentrate disposal and water recovery from RO technologies is to run a system with zero liquid discharge, or ZLD, where the water recovery would approach 100% [8]. In ZLD, water is completely separated from the salts in the concentrate. To achieve ZLD some common methods such as mechanical vapor compression brine concentration and evaporation pond crystallization and membrane-based emerging methods such as multi-stage RO with pre and post-treatment, forward osmosis (FO), electro-dialysis (ED), electro-dialysis reversal (EDR) and membrane distillations (MD) are used [14]. The benefits of ZLD systems include its flexibility of operation in any geographical location, more acceptable by local communities as they are minimum waste generation systems and have positive environmental effects. However, the major issues with a ZLD system are its high energy demands and operation and maintenance cost to achieve 100% recovery [8]. One feasible strategy to recover costs is via selective recovery of valuable constituents in RO concentrate. The recovery of these constituents can potentially be marketed to increase the profitability of desalination units. Additionally, recovered materials can be potentially reused within the desalination facilities to bring down the operational costs. A study estimates that mining sodium (Na), potassium (K), magnesium (Mg), calcium (Ca), strontium (Sr), chloride (Cl), bromine (Br) and rubidium (Rb) could be financially profitable and can potentially make a desalination unit revenue generator [15]. Moreover, the recovery of these ions from RO concentrate would be more feasible because concentrations are approximately double that of feed seawater. The salts valued at 895, 000 USD per annum can be profitably recovered from 405,000 m^3 of RO concentrate every year [10]. Going by the projection of increasing desalination production capacity worldwide (by 2030 approximately 1.2 x 10^{11} tonnes water production per year [16]), it is imperative to recover useful resources from RO concentrates both from environmental and economic perspectives.

In this book chapter, we will attempt to discuss the RO-based desalination process briefly, its theoretical principle, process efficiency, and estimate the resource recovery and water reuse aspects from RO concentrate and technologies that are currently employed based on the literature available so far.

2. RO DESALINATION

2.1. Basic Theoretical Principles and Operation

Membrane filtration schemes are broadly classified as nanofiltration (NF), ultrafiltration (UF) and microfiltration (MF) and RO. Out of these, RO membranes are designed to reject the smallest contaminants, i.e., monovalent ions, whereas NF, UF, and MF reject, species of greater size. The nominal pore diameter of RO membranes is in the range of 1 – 10 Å. In RO membranes, polymeric material forms a layer, through which water flows in a tortuous path and come out as permeate. RO membranes are commercially available as spiral wound modules, in which membrane sheets are wound around an inner tube where the permeate is collected. In typical membrane filtration, a positive hydrostatic pressure is applied to force the fluid through pores. The permeate rate is a function of membrane porosity and tortuosity; diffusional aspects are also important in determining the overall flow rate [8]. Mathematically, the permeate flux of a component ‘i’, can be expressed as [17]:

$$N_{ix} = \frac{\rho_i k}{\mu}\frac{dP}{dx} - D_{ij}\frac{d\rho_i}{dx} \quad (1)$$

Where N_{ix} is the permeate flux of i in the x-direction, ρ_i is the density of i, k is membrane’s permeability, μ is the fluid viscosity, $\frac{dP}{dx}$ is the pressure gradient in the x-direction and D_{ij} is the binary diffusion coefficient of i with respect to j within the membrane. The transport mechanism in RO is primarily diffusion-controlled and has been termed as solution-diffusion model [8]. Respective three major steps involved are 1) absorption of water molecule over the membrane surface 2) subsequent diffusion (due to concentration gradient) through the membrane thickness and 3) desorption of water molecule from the permeate side of the membrane into the bulk permeate [18, 19]. In RO systems, hydrostatic pressure greater than the osmotic pressure is applied to create a positive pressure difference which subsequently leads to the creation of chemical potential difference (concentration gradient). This gradient drives the water movement against the natural direction of osmosis, i.e., water movement from denser (high concentration) medium to rarer (lower concentration) medium, and traps the salt on the influent side of the membrane surface, leading to the formation of RO concentrate solution on end, and almost salt-free solution as permeate at another end of the membrane. In terms of osmotic pressure ($\Delta\Pi$) and transmembrane pressure (Δp) differences across the membrane, water flux is expressed as:

$$N_i = L(\Delta p - \Delta\Pi) \tag{2}$$

Where L = permeability coefficient, which depends on the membrane properties such as water diffusivity and solubility within the membrane, water partial molar volume, membrane thickness and operating temperature [17].

The osmotic pressure is a colligative property of the fluid system which depends upon the total concentration of the ionic species in the fluid, and it is expressed as [20]:

$$\Pi = iCRT \tag{3}$$

Where i = van't Hoff index; C = molar concentration of the solute; R = universal gas constant and T is the operating temperature.

For a typical seawater sample, the osmotic pressure is around 2300 – 2600 kPa and sometimes can reach up to the value of 3500 kPa; for brackish water, the typical range is 100 – 300 kPa for the concentration range of 2 – 5 g/L [8]. The recovery of water (R) by the RO process is strongly affected by this parameter, and the dependence is expressed as follows [21]:

$$\Pi_{concentrate} = \frac{1}{1-R} \quad (4)$$

The recovery can also be expressed as the ratio of permeate and feed flow rates. The typical recovery rate observed is in the range of 35 – 85% [8]; for seawater desalination, the corresponding figure is 45 – 55% [11]. To improve the recovery rate, the osmotic pressure needs to be overcome by the feed pressure, which for seawater applications range between 6000 – 8000 kPa [8]. The key barriers to improving process performance are increasing osmotic pressure on the concentrate side, fouling potential of the feed and mineral scaling. Moreover, concentration polarization significantly reduces process performance. Concentration polarization occurs when salts (the retained ions) begin to deposit over the feed side of the membrane, leading to the formation of a layer; this increases the local osmotic pressure at the membrane surface, and thereby reduces the overall driving force (i.e., difference between hydrostatic pressure and osmotic pressure) and affecting the recovery rate. The salt permeation increases through the membrane and consequently can lead to the phenomena of salt precipitation and fouling on the membrane surface. The salt rejection is commercially considered an important performance parameter for RO membranes. The salt rejection R_s is expressed as

$$R_s = \left(1-\frac{C_{feed}}{C_{concentrate}}\right)*100\% \quad (5)$$

Both R_s and R determine as to what would be the concentration of salts in the concentrate. With increasing recovery rate (R), it has been reported that concentrate total dissolved solute (TDS) increases exponentially and affects the membrane performance. Therefore, limited options exist to increase the recovery rate by once-through membrane operations, and often to improve the recovery rate a series/parallel arrangement of RO is applied where permeate of one is fed to the next RO in series. In typical desalination plants, 1 – 4 passes are applied (permeate is fed to the next in the chain) or stages (concentrate is fed to the next in the chain) [22]. The RO concentrate that is finally generated from the system is then discharged as waste.

The available evidence suggests that advancement in RO membrane technology is critical to increasing the water reuse projects worldwide. Relevant examples include the improvement in water reuse capacity in Orange County Water District in California which began in 1970s, which has been utilizing RO membrane desalination and increasing the wastewater reclamation by producing potable drinking water in the range of 70 million gallons per day (MGD) in 2008 to 100 MGD in 2015 with a target to reach a capacity of 130 MGD by 2023 [23]. However, the key issues plaguing RO membrane processes and preventing its further growth is its high energy requirements, membrane fouling, and RO concentrate disposal. These issues have been addressed by progress in membrane development and process innovation.

Membrane development implies the utilization of novel materials or nanomaterials for membrane fabrication. The examples include thin-film composite (TFC) polyamide membranes [24], thin-film nanocomposite (TFN) membrane [25], nanostructured polyamide membrane and membranes made up of next-generation materials such as aquaporins, graphene, carbon nanotubes (CNT) and metal-organic framework embedded membranes [26]. The reported results show improved and efficient membrane performance with respect to water flux and membrane area requirements. However, most of these technologies are currently in bench scale to pilot scale, with only TFC/TFN based membrane processes have applied for a full-scale operation. Moreover, TFC membranes are prone to fouling, and their fabrication technique has not been extended to hollow fiber configurations which give more compact packing densities. The promising results in terms of higher water flux shown by the materials such as carbon nanotubes, aquaporins, etc. still need some research regarding their energy requirements and redesign of membrane modules to utilize them fully. Process innovation includes the development of FO and/or hybrid processes such as FO and RO together [27] or with MD with crystallization [28]. Almost all these hybrid processes presently are on pilot scale. One another way of process innovation can be achieved by integrating RO concentrate treatment and reuse in-line with RO processes. This has the potential of improving the economics of the overall process by incorporating resource recovery as well as of improving the water recovery by recycling the TDS-reduced water in the feed cycle. In the subsequent sections, the feasibility of incorporating RO concentrate treatment in the RO process will be discussed with respect to the assessment of the available state of technology and potential revenue generation.

3. RO Concentrate Management

3.1. RO Concentrate: Composition and Recoverable Species

RO concentrate is mostly the concentrated form of the respective feed water used in RO processes. Therefore, in seawater desalination, RO concentrate has the same components as seawater in addition to the pre-treatment chemicals such as antifoulants, antiscalants, coagulants and cleaning chemicals which include surfactants, alkaline, acid solutions, and metal chelating agents. However, the large proportion is of seawater components. The main ions which constitute 99.9% of the salts in seawater in decreasing order of their concentration are Na^+, Mg^{2+}, Ca^{2+}, K^+ and Sr^{2+} for cations and for anions the species in decreasing order of their concentration are Cl^-, SO_4^{2-}, HCO_3^-, Br^-, BO_3^{2-} and F^- [16]. The typical RO concentrate composition is provided in Table 1. Traditionally, salts of Na, Ca, Mg and K have been commercially extracted from seawater in the form of chloride, sulfate, carbonate, and oxide. However, recent advancements and proliferation of desalination units worldwide have opened the potential of extracting other minerals from the concentrated discharge solutions. A study estimates that in a desalination plant in Saudi Arabia can yield a potential annual revenue of $ 18.909 billion by recovering only Na, Ca, Mg and K from RO concentrate [29]. With decreasing cost of desalination and rising production of desalinated water, it is hypothesized that other species (lithium (Li), rubidium (Rb), strontium (Sr), iodine (I), bromine (Br), boron

(B)) would also be recovered profitably in the near future. According to Loganathan et al. (2017) [16], Na, Ca, Mg, K, Li, Sr, Br, and uranium (U) are potentially profitable for extraction, provided suitable methods of recovery are developed that are more economical than mining these elements from the land. Shahmansouri et al. (2015) [15] after studying the economic feasibility of recovering different salts from seawater and RO concentrate using a profitability model that included market price, the concentration of the element, RO concentrate flow rate and operation and maintenance cost, reported a similar list of elements. The additional elements in the list included cesium (Cs), silicon (Si), and chlorine (Cl). Some of the elements (Li, molybdenum (Mo), N) are borderline cases whose profitability depended upon the flow rate of RO concentrate. Other elements included in the study which can improve the economics of desalination are cobalt (Co), germanium (Ge), and phosphorus (P). Alternative recovery such as chlorine for applications (e.g., NaOCl as a disinfectant [30]) within the desalination facilities can offset some desalination costs. Roughly 4.5×10^9 tons of U is presently distributed in seawater out of 1.37×10^{18} m^3 total volume of oceans. Currently, U caters to 16% of our electricity needs and is expected to increase [31]. Therefore, extracting U from seawater can serve as a potentially inexhaustible supply source. Salt removal also offers an opportunity of recycling the treated water which would have reduced TDS as feed.

This would translate into additional water recovery and augmenting the water supply and reduce the cost of water. There is also demand from industries and agriculture for minerals which are available in seawater in much greater quantities than available on land [16]. Therefore, it is imperative to assess the current technological development if it can be utilized for resource recovery applications from RO concentrate where profitability would chiefly depend on target components' concentration in RO concentrate, their market prices and volume of RO concentrate produced. It is also important to review technology based on their stage of development and feasibility of achieving resource recovery targets from RO concentrate.

3.2. RO Concentrate Resource Recovery Technology

Traditionally, RO concentrate waste management strategies have largely been dealt with either by direct disposal or by minimizing their volume produced. Disposal methods include inland and offshore strategies. The minimization schemes have been carried out by emerging membrane and thermal-based processes. Another strategy, which is more sustainable, is the reuse of reject brine for useful mineral recovery. There are different technologies available that recover salt and chemicals to minimize the overall cost of desalination. Recovering salts from RO concentrate not only bring economic benefits but also reduces the overall environmental footprints of the desalination process.

3.2.1. Chemical Precipitation

Chemical precipitation has been an application for some time for salt removal from seawater and RO concentrates. Mg and Ca were precipitated using sodium carbonate and sodium phosphate as precipitants from the RO concentrate generated in desalination units located around the shores of the Red Sea and the Mediterranean Sea. Using sodium phosphate, Ca and Mg recoveries were ~ 75% and ~ 24% respectively. However, with sodium carbonate Ca recovery was ~ 95% and Mg ~ 9% [32]. The salts are then subsequently removed via a solid-liquid separation step such as sedimentation and filtration. The process is simple and governed by ionic strength and pH of the solution. The precipitation step also lowers the mineral scaling tendency of the RO concentrate. Therefore, it has chances of improving the water recovery rate in the subsequent secondary RO process and effectively contributing to water reuse and minimization of RO concentrate volume. The overall Ca recovery was 654 mg and for Mg recovery was 225 mg per liter of RO concentrate treated. The key drawbacks mainly are the use of additional chemical reagents, which adds to the cost, the additional steps to separate the desirable product from the solution, and also maintaining the purity of the products is challenging since during precipitation, side precipitation reaction or seeding crystallization of other species can occur over the solid precipitate phase. Sludge handling and disposal and accurately maintaining the system pH also add to the cost and environmental impact of the process. Chemical precipitation is mainly used in conjunction with another process such as NF filtration or ion-exchange process to improve the yield and purity of the product which also adds to the overall cost [33]. Overall as a step in a multiple RO process, intermediate chemical precipitation removes scale-forming minerals such as $CaSO_4$ and therefore improves water recovery rate in secondary RO and thus leads to greater freshwater availability and environmental viability. For continuously run processes, retrofitting, optimization and economic analysis is important. A study reports the capital cost for intermediate chemical softening approximately $ 5511 per m^3 of RO concentrate per hour. The reported chemical cost is around $ 2.72 per m^3 of RO concentrate [14].

Another alternative precipitation method is seeded precipitation. In this method, RO concentrate which has certain species in supersaturated state (which are sometimes stabilized by the carryover antiscalants) is inoculated with seed crystals to enable concentrate desupersaturation via precipitative seed crystal growth. This method requires less chemicals than direct precipitation because it utilizes the supersaturation of some species to drive crystal growth and add only seeding agents to make the supersaturated species to grow as crystals [34]. Approximately 50% of the drinking water in the Netherlands has been softened by the seeded precipitation method where pure calcite pellets are produced for reuse with $CaCO_3$ or sand as seeding materials [35]. The main drawback to this process comes from the carryover of antiscalants in the RO concentrate which interacts with the supersaturated species and can arrest the crystal growth in seeded precipitation. The possible solution to prevent the interference of antiscalants is by using some organic chelating agents, polymeric coagulants or by their chemical/electrochemical oxidation. However, this approach has an issue of carryover of these additives into the next stage RO process feed, which can impact the fouling level in the subsequent RO processes [36, 37]. Rehardianto et al. (2010) [34] came up with a sequential scheme known as chemically enhanced seeded precipitation (CESP), wherein a stepwise approach, antiscalants are first removed in limited precipitation of $CaCO_3$ by using a small amount of alkaline chemicals (lime, NaOH, Na_2CO_3).

Antiscalants are trapped within by adsorption on the solid matrix of precipitating $CaCO_3$ surface overgrowth and aggregation. This eliminates antiscalants without introducing any foreign additives. Subsequently, the remaining Ca is seed crystallized and recovered as gypsum ($CaSO_4$). Relative to chemical precipitation, CESP utilizes the alkaline reagents minimally, to remove the antiscalants from the solution and thereby precluding the entry of foreign additives that may disrupt the RO process downstream. The drawback is that antiscalants can themselves retard the nucleation and surface growth of $CaCO_3$ particles and therefore, its overall precipitation. Though alkaline dose is minimal, the rate of $CaCO_3$ precipitation and the extent of antiscalant removal are governed by its dose. The alkaline dose indirectly controls the crystallization of gypsum in the next step. This is because during seed crystallization via gypsum seed crystals, adsorption of remaining antiscalants is an important step to enhance nucleation as these antiscalants provide surface area for epitaxial surface growth. Therefore, the alkaline dose remains an important element in the overall CESP process. Rahardianto et al. (2010) [34] showed utilizing CESP with gypsum as seeding crystals at a dose of 2.5 g/L or 4 g/L improvement of water recovery rate from 63% to 87%, with gypsum and $CaCO_3$ as final recovered marketable commodities. Xiong et al. (2017) [38] applied seeded precipitation on RO concentrate of coal chemical wastewater and achieved 90% water recovery and substantial gypsum crystals. Pervov (2015) [39] utilized an open channel membrane module to carry out seed crystallization of $CaCO_3$ which increased the water recovery to 95%. A three-step method was developed by Aghdam et al. (2016) [40], wherein the first step $FeCl_3$ was used to remove phosphonate-based antiscalant, the second step involved pH adjustment to alkaline range to cause carbonate precipitation and finally in the third barium sulfate was seed crystallized using barite seed crystals; the study reported > 97% water recovery from the overall process. Pellet softening using $CaCO_3$ seed crystals in a fluidized bed can potentially be utilized for both water and resource recovery since in this method, dry sludge is produced which can be easily separated and further treated for resource recovery [41]. The use of the pellet reactor has shown enhanced water recovery from the RO process (from 80% to 90%) in Saudi Arabia's Qasim inland desalination units [41]. However, pellet softening is not as effective when it comes to removing Mg ions, due to the latter's higher solubility than Ca [33]. This is useful from a selective resource recovery perspective and can be utilized in cases where Mg and Ca are to be recovered separately. Another similar technique is seeded slurry precipitation (SPARRO), where unlike pellets, seed crystals are in the form of a slurry, which attracts nucleation of calcium sulfate and silicone dioxide [33]. It is mainly utilized for RO concentrate rich in calcium sulfate.

One main limitation of chemical precipitation/seed crystallization is that it requires a high level of supersaturation of the precipitating species. It is limited to extracting only Ca and Mg salts as pre-treatment of feed to primary and secondary RO process and thereby reducing scale forming propensity and improving the water recovery level. It has limited applications with respect to the other undersaturated or trace ions that can be recovered from RO concentrate. Therefore, to tap into the remaining resources from RO concentrate, a more encompassing technological alternative would be more suitable.

Table 1. Composition and physical properties of Mediterranean seawater and RO concentrate [50]

Parameter	Mediterranean Seawater	RO Concentrate
Ca^{2+} (mg/L)	400	690
Mg^{2+} (mg/L)	1460	2600
Na^{+} (mg/L)	11000	24000
Cl^{-} (mg/L)	20000	40000
SO_4^{2-} (mg/L)	2700	5400
K^{+} (mg/L)	500	800
CO_3^{2-} (mg/L)	20	43
HCO_3^{-} (mg/L)	90	146
Alkalinity (mg/L)	110	190
TDS (mg/L)	36320	73870
Conductivity (μs)	57500	108700
pH	7.6	7.87

3.2.2. Adsorption and Ion-Exchange Based Processes

Since RO concentrate is a rich resource for metals, which are mostly in ionic forms and trace concentration levels, techniques like adsorption and ion-exchange can be useful in selectively removing and recovering these trace or low concentration metal species in the presence of high levels of ions such as Ca^{2+} and Mg^{2+}. Adsorbents/ion-exchangers have certain functional groups, which are either naturally present or grafted synthetically, and have high selectivity for the targeted species. Therefore, choice of adsorbents/ion-exchangers primarily depends upon the physical and chemical properties of the metal ions, which in turn requires that sorbents have high uptake capacity and selectivity for the targeted metal ions against the background competing ions which are in much higher concentration. After the sorbent is exhausted, there is an additional step of desorption where the targeted species is quantitatively removed from the sorbent using a minimum volume of regenerant solution [42]. Further, the concentrated solution of regenerant and targeted species is treated to obtain the desired product. The additional steps of regeneration and final recovery from regenerant may add to the processing cost, but since adsorbent or ion-exchanger and regenerant can be reused in multiple cycles, the overall cost over the lifecycle of the process can be potentially reduced. The reuse and recycle of the raw materials cut down the material requirement and lower the environmental footprints of the process and make it sustainable following the 3R principle of sustainability i.e., reduce, reuse and recycle.

The selective extraction of minerals from RO concentrate using sorbents is more feasible than direct extraction from seawater, because RO process increases the ionic strength of seawater by more than twice or thrice. However, there is also a greater tendency of the targeted ions getting screened by other competing ions.

The adsorbents can be natural or synthetic. Natural sorbents are typically low-cost materials and may come from agricultural sources (fruits, vegetables, food), by-products, residues and wastes, minerals, rocks (zeolites), and marine organisms' shells (alginate, chitosan), etc. [43]. The application of natural waste materials as adsorbents further makes the process green and environmentally friendly. Synthetic sorbents include organic, inorganic and polymeric, chelating ion exchange resins and nanomaterials. Synthetic sorbents are costlier than natural sorbents but usually have higher uptake capacity and selectivity. The key mechanism governing selective adsorption of a species by a functional group is by ligand exchange, chelation, inner-sphere complexation or exchange of ions which have high mutual affinity arising out of their acid-base nature based on hard-soft acid-base (HSAB) theory given by Pearson. The use of adsorption and ion-exchange for mineral recovery has largely been in laboratory-scale studies [16].

The recovery of Li, Cs, U, Rb, Sr, Ca and Mg using adsorbents has been reported in literature, out of which the most studied minerals are Li, Rb, Sr, and U. Li was mainly extracted using a MnO_2 – based adsorbent converted into H form or λ – MnO_2 [44, 45]. Chitrakar et al. (2001) [46] studied 12 inorganic adsorbents for selectively removing Li from seawater, and reported that H – form of MnO_2 has the highest uptake capacity; the selectivity of H-MnO_2 was 5 – 6 order of magnitude higher than the other cations (Na, K, Mg and Ca) present in the system with maximum uptake capacity of 40 mg Li per g of the adsorbent. Li was also recovered in a packed-bed column of λ-MnO_2 by passing seawater as influent. The packed bed column selectively concentrated Li from 0.16 mg/l in seawater to 900 mg/l in the eluent of HCl; further, Li was recovered as 99.9% pure Li_2CO_3 [44]. In a study utilizing commercial sorbents (CsTreat (hexa-cyannoferrate based sorbent), ZrP (zirconium phosphate) and S910 (amido-oxime functionalized resin)), consortium of metal ions (Cs(I), Li(I), Rb(I) and U(VI)) (concentration of target ions spiked to 20 mg/l) were removed from RO concentrate obtained from El Prat RO plant in a packed-bed column mode.

The results show that CsTreat achieved the highest uptake capacity for Rb(I), with fairly high level of Cs(I) uptake. S910 showed good uptake of U(VI). However, Li retention in ZrP was poor, mainly due to the competition by Na^+ ions in RO concentrate [47]. For trace metals to be effectively removed by ion-exchange or adsorption process, it is necessary that the kinetics should be fast, and target metal ions should be instantaneously sorbed when they contact the ion-exchanger. The key to improving the kinetics lies in enhancing the diffusion of metal ions through the sorbent, which can be achieved by modifying the role of the cross-linking agent, its nature or density on the swelling properties of the resin. The study noted that to get an effective removal at the unspiked level of the metal ions would be a challenge [47].

Extracting U can make a significant difference in power production. Since the concentration of U in seawater is trace, therefore, developing a selective sorptive process becomes more necessary. Among inorganic resins, hydrous titanium oxide and lead sulfide show high uptake capacity; organic sorbents such as Calix [6] arenes, polyacryamidoxime, polyhydroxamic acid, and resorcinol arsenic acid have shown high uptake capacity. Of these functional groups, amidoxime group shows the highest adsorption capacity (3.9 mg U per g of the sorbent) and stronger selectivity for U in the presence of alkali and alkaline earth metal ions [31]. Pal et al. (2010) [31] synthesized cross-linked polyacrylamide and converted it to polyhydroxamic acid resin to recover U from 3.6 and 4 ppb U solution (which is the typical concentration for seawater brine streams). The resin also displayed high selectivity for U in the presence of Ti, V, Co and Mo in a column mode. However, the study did not consider the other high background concentration species in the RO concentrate and their impact on U sorption. The resin is suitable for RO concentrate with a slightly acidic pH. A highly porous metal-organic framework bearing an orthogonal phosphorylurea group in its chain demonstrated an uptake capacity of 188 mg U per g of sorbent (as compared to 54 mg U per g for amidoxime resin), which was attributed to the binding of uranyl ions with two phosphorylurea ligands [48].

In a study done on artificial brine, it is noted that the key U species present in the brine around neutral pH is $UO_2(CO_3)$. The most important step in adsorption of U by amidoxime group is the decomposition of $UO_2(CO_3)^{4-}$ to UO_2^{2+} which is then followed by complexation of UO_2^{2+} by four amidoxime groups [47]. U(VI) is also usually removed by phosphonic and sulfonic-based resin.

These groups have shown higher adsorption capacity (22.8 mg U per g as compared to 5 mg U per g of amidoxime group), mainly due to the presence of negatively charged bifunctional phosphonic/sulfonic groups which strongly attracts the positively charge UO_2^{2+} cations [49].

Table 2. Summary of RO concentrate treatment methods

Method	Advantages	Limitations	Key results
Chemical precipitation	Lowers mineral scaling in secondary RO Improves water recovery rate Minimizes RO concentrate volume	Requires additional chemical reagents and separation steps Adds to the overall cost Maintaining the purity of the product is a challenge Useful only for species in high concentration, supersaturation, and low solubility	95% & 75% Ca and 24% and 9% Mg recovery by direct precipitation with Na_3PO_4 & Na_2CO_3 respectively [38] Sequential recovery of $CaCO_3$ & $CaSO_4$ and 87% water recovery using CESP [40] Seed precipitation of $CaSO_4$ & 90% water recovery [43] $CaCO_3$ & 95% water recovery [44] $CaCO_3$, $BaSO_4$ & 97% water recovery [45] Pellet softening; $CaCO_3$ & 90% water recovery [46]
Adsorption/ion -exchange	Useful for selective recovery of trace/low concentration ions against high background concentration ions Low cost, simple operation & maintenance Can be used multiple times after exhaustion & regeneration Potentially lucrative as high market value minerals can be recovered	Chance of parallel sorption of high concentration background species, and thereby affecting the purity of final products Limited scope as a standalone RO concentrate treatment method; requires integration with other processes	Li recovered from seawater using H-form MnO2 with uptake capacity of 40 mg Li/g and selectivity 5-6 order of magnitude higher than Na, K, Mg and Ca [51] Li recovered as 99.9% pure Li_2CO_3; concentrating Li from 0.16 mg/L in the feed seawater to 900 mg/L in the regenerant using a packed-bed column [49] Cs, Li, Rb, U recovered using Cs-treat, ZrP & S910 adsorbents [52] Amidoxime group selectively recovered 3.6 – 4 ppb U from RO concentrate [37] U recovered by phosphoryl urea group-based sorbent at an uptake capacity of 188 mg U/g [53] Ca^{2+}, Mg^{2+}, Sr^{2+} (divalent cations) recovered by alginate hydrogels [55, 56] Potassium hexacyanoferrate recovered Rb with highest uptake capacity of 47 mg/g (batch) & 238 mg/g (column) [52, 54]

Method	Advantages	Limitations	Key results
Evaporative methods	Simple to design and operate, low maintenance Economic for recovery of salts such as NaCl, KCl, $CaSO_4$, $MgSO_4$, $CaCO_3$ Useful for ZLD Suitable for arid areas	Large land area requirement, longer duration Prone to cause soil and groundwater pollution Water recovery and reuse not possible Controlled environment needed to dictate the sequence of salt precipitation	Sequential recovery of NaCl, KCl, $CaSO_4.2H_2O$ & $MgSO_4.7H_2O$ from a hexary system of Na^+, Mg^{2+}, K^+, Ca^{2+}, Cl^-, SO_4^{2-} [62] Heat exchanger integrated solar & evaporation pond with an increased evaporation rate [65] WAIV induced rapid evaporation [66]
MDC & ED	Desired crystal morphology with low CVs Integrating with RO/NF processes can lead to lower cost and increased water recovery More cost effective than evaporators and brine evaporation ED separates monovalent and divalent cations; product purity high for salts like NaCl	Only practical for minerals which are in higher concentration and supersaturation Polarization and increased resistance to vapor flow at high water recovery rate ED prone to fouling	NaCl crystals recovered with low CVs [72] MgSO4 crystals with CVs between 0.31-0.41 [73] 100% water reuse with lower cost than standalone RO [76] Li ion specific ED process concentrated Li ions from seawater [79]

Hydrogels made up of alginates, and chitosan have also been utilized for removing divalent cations such as Ca^{2+}, $Mg^{2+,}$ and Sr^{2+} in seawater brine by sorption [50, 51]. The uptake capacity reported was 54 mg/g and 34 mg/g for Ca^{2+} using alginate and chitosan hydrogels respectively; for Mg^{2+} the figures were 280 and 316 mg/g for alginate and chitosan respectively [50]; for Sr^{2+} cations, the corresponding value for alginate hydrogel was 110 mg per g [51]. The higher uptake capacities of the hydrogels for Mg and Sr are mainly due to the higher selectivity for Mg^{2+} and Sr^{2+} as compared to Ca^{2+}. The alginate and chitosan are natural and biodegradable polymers that are cheaply available; secondly, they can be pretreated by a variety of methods like grafting other functional groups such as acrylamide, microwave, UV treatment to increase their uptake capacity and selectivity [50]. The sorption of divalent cations by alginate is mainly due to the cross-linking of guluronic acid groups by divalent cations, which acts as a bridging species [52]. In the presence of Ca and Mg cations, it is noted that Sr uptake capacity is reduced mainly because of the competition by Ca and Mg ions to act as a bridging species [53]. Another adsorbent for Sr is titanium nanotube, which has shown a high uptake capacity of 92 mg Sr per g of the sorbent [54]. Ca also was noted to hamper Sr uptake by titanium nanotube. Subsequently, removing Ca ions by precipitation (as $Ca(OH)_2$ by adding NaOH since $Ca(OH)_2$ has two order of magnitude lower solubility product constant (K_{sp}) than $Sr(OH)_2$ (K_{sp} $Ca(OH)_2$: 5.5 x 10^{-6} and K_{sp} $Sr(OH)_2$: 3.2 x 10^{-4} [55])) before introducing the sorbent increased its Sr removal capacity. The mechanism involved Sr exchange with Na in the interlayer of titanium oxide octahedra [54]. For Rb, potassium cobalt hexacyanoferrate (KCoFC) has been reported to show the highest uptake capacity with capacities of 47 mg per g and 238 mg per g in batch and column runs [47, 49]. The uptake capacity in another study followed the order as Rb > Cs > Li > Na, Ca [56]. The main reasons for this trend were the high charge density of Rb due to its lower hydrated ionic radius and due to the deeper penetration into the crystal lattice to replace K from the center of KCoFC; also since Rb and K have similar unhydrated ionic radii, Rb is better able to replace K from the crystal lattice among other cations present in seawater. Naidu et al. (2016) [57] studied different metal forms of KCoFC by replacing Co with Cu, Ni, and

Zn, and reported that Cu form showed the highest sorption capacity for Rb. The presence of other cations (Na, Ca, Mg and K) was also studied, and it is seen that all other cations reduced the sorption capacity for Rb, but K reduced it remarkably. Therefore, 0.1 M KCl is chosen for regenerating the packed-bed column used for the study and 95% regeneration was achieved.

The key advantages of using the adsorption system are its low cost, simple operation and low maintenance. Moreover, the sorbents can be used in multiple cycles upon regeneration which also greatly reduces the overall operating and capital cost. From a process perspective, it is most useful in the case where selective recovery of trace species is required against high background ionic strength of the solution. Since in RO concentrate background concentration is considerably high, therefore it is imperative to develop sorbents with such functional groups that offer remarkably high selectivity and separation factors. One downside to the adsorption process in the context of RO concentrate is the purity of the final product since other components will also be simultaneously adsorbed by the sorbents, which would desorb with the target metal ions in the eluting solution. Secondly, adsorption/ion-exchange processes can never be a stand-alone RO concentrate treatment operation since it would not significantly affect the TDS of RO concentrate and thereby water recovery rate. Therefore, adsorption-based processes can only be used in conjunction with other processes. From a profitability perspective, adsorption-based processes have high potential since it can recover high market value minerals, and it can bring down the overall operating cost of desalination process plants.

3.2.3. Crystallization and Evaporation-Based Methods

The crystallization-based approach to salt recovery is termed as selective salt recovery (SSR) [33]. By changing temperature, or using an agent, salts are selectively recovered via precipitation induced by a change in salt concentration. The tendency to co-precipitate/adsorb and solubility differences of the salts determine the overall operation sequence for salts precipitation and their purity. Hajbi et al. (2010) [58] studied the solubility diagram and crystallization paths of a hexary system consisting of Na^+, K^+, Mg^{2+}, Ca^{2+}, Cl^- and SO_4^{2-}. The results show the feasibility of recovering salts

(NaCl, KCl, $CaSO_4.2H_2O$, $MgSO_4.7H_2O$) by a combination of isothermal and isobaric evaporation of the rejected brine. The process is influenced by operating conditions such as pH, pressure, and co-components apart from targeted components. The method has economic viability and potential for mitigating environmental problems caused by disposing of the RO concentrate. However, the high capital cost limits the process application. For a full-scale brackish water RO system with 75 – 85% water recovery the energy consumption for brine concentrator is in the range of 18 – 26 kWh/m^3 with \$ 33,000 per m^3 per h as the capital cost and \$ 0.01 per m^3 as the chemical cost [14]. SAL-PROC™ and brine concentration are two of the most common SSR approaches yielding products which are marketable and of commercial grade. SAL-PROC™ process can recover marketable salts such as gypsum, limestones, Mg $(OH)_2$, sodium sulfate, sodium chloride, and calcium chloride. The process requires a controlled process sequence and evaporative cooling, which can add to the cost and limit its application. Moreover, water recovery and reuse are not possible through this approach. However, in cases where ZLD is required, it may be applied, with only salt recovery as the incentive [59].

Evaporative methods involve natural evaporation of RO concentrate using solar energy. The RO brine is gradually concentrated with salts in a large pit. It is cost-effective and most suited for arid areas, as land there will be available with no danger of contaminating groundwater from the leakage of minerals. The pits are divided into three pans, the first one receiving pan, followed by evaporator pan and finally crystallizer pan. The salts precipitate in the order of their decreasing saturation index, as predicted by thermodynamics, therefore, in evaporator pan, first $CaCO_3$ precipitates and then gypsum. In the crystallizer pan then NaCl crystallizes when the TDS is highest [16]. Most of these evaporative techniques have been applied directly to seawater to recover salts. The evaporative ponds are simple to design and have low maintenance and operating requirements and cost. However, they require a large land area (an area of 700, 000 m^2 required for 5000 m^3 per day of concentrate in Sabha RO desalination plant near the Red Sea port of Eilat), longer duration (low evaporation rate, roughly 4 L/m^2-day) and are prone to land pollution [10]. Sometimes ponds are lined with clay, polyvinyl and polyethylene sheets/coatings/ liners or made with galvanized iron [60, 61]. Additionally, the process fails to recover water. All these factors increase the environmental footprints of the process. Some recent developments include the use of heat exchangers between the solar pond and evaporative pond to enhance the evaporation rate for desalination brines [61]. Another development is the utilization of wind energy to enhance the evaporation by a method called Wind-Aided Intensification of eVaporation (WAIV), wind energy is utilized to evaporate water from the brine-wetted fabrics [62]. The hydrophilic fabric is vertically arranged in a high-density packing and placed parallel to the wind direction. WAIV requires less land and have shown higher and faster evaporation than solar ponds. The hydrophilicity of the fabric is important as too hydrophilic fabric would reduce the effective vapor pressure of the surface. The evaporative techniques are cost-effective but require large land. It is economically useful only for minerals with high concentration.

3.2.4. Membrane-Based Methods

The membrane processes developed/reported for RO concentrate treatment typically consist of membrane distillation, electrodialysis, and FO. NF has also been reported for the removal of sulfate and chloride from the brine where 75-96% of sulfate and 2-11% of chloride was rejected at 500-2000 kPa for the brine with ionic strength ranging as 0.4 – 1.3 M [64]. However, resource recovery potential from NF is not significant, as it further adds to the generation of reject brine, which then requires treatment and salt separation for final recovery. Therefore, in this section, we will keep our discussion to the first two schemes described above, which are more important from a resource recovery perspective.

Membrane distillation is an evaporative process wherein a hydrophobic microporous membrane is utilized to separate pure water as a distillate from the brine solution. The hydrophobic property of the polymeric membrane helps in keeping out the water molecules from the pores by creating a vapor/liquid interface at the entrance of each pore. Due to thermal evaporation (operating temperature range: 60 - 80°C), the water vapor generated at the interface on the retentate side (warm side of the membrane) diffuse through the pores and condenses on the distillate side (cold side of the membrane) [64]. The key advantages of membrane distillation over conventional distillation are its lower operating temperature, high specific surface area, making the set-up compact and with a smaller footprint. Membrane distillation has been demonstrated to yield a higher water recovery rate. However, its ability to recover salts is limited. These factors make membrane distillation high on environmental sustainability but low with respect to economic benefits. Therefore, a variant, known as membrane distillation crystallization (MDC), has been developed that facilitates salts recovery by crystallization. The MDC process allows the concentration of salts up to the supersaturated state, which then subsequently leads to their crystallization. In MDC, a hydrophobic porous hollow fiber membrane is used to maintain a controlled supersaturated environment where crystallization occurs in a circulating crystallizer, and finally, recovery is carried out in a crystal recovery system, which prevents their deposition within the MDC plant [65, 66]. The process leads to a metastable state where crystal nucleation and growth take place [67]. The high surface area provided by hollow fiber membranes helps in achieving reasonable evaporation flux at a moderate temperature level of 40 - 50°C with energy expenditure of about 15 – 20 kWh/m^3; comparatively a conventional evaporative NaCl crystallizer operating at 70°C takes about 30 kWh/m^3 [10].

In MDC, since the process environment with respect to crystal growth is more constrained and controlled in hollow fibers than conventional crystallizers, therefore crystals of desired structure and morphology can be obtained. The axial flux of the homogenous diffusing solution through the hollow fibers provides uniform orientation to the nucleating crystals and lower mechanical stress, therefore it leads to a symmetric environment and results in a uniform size distribution [16]. NaCl crystals with low crystal diameter coefficient of variation, CV, (i.e., size distribution in a narrow range) were obtained from synthetic seawater RO concentrates [68]. Similarly, $MgSO_4$ crystals with low CV between 31 – 41% were produced from RO concentrate. MDC based methods have largely been operated on a laboratory scale and not yet commercially present on an industrial scale.

Studies have reported the method potential in extracting minerals such as Ba, Sr, Li, Cu and Ni from RO and NF concentrate [69].

Table 3. Summary of key minerals recovered, their scale of recovery and market price

Element/Salt	Scale: Method	Market Value ($/kg)*
$CaSO_4$	**Lab-scale:** 1. Chemical precipitation using Na_2CO_3 and on-site RO concentrate [38] 2. CESP using lime dosing in the first step followed by gypsum seeding [40] **Full-scale:** 1. Pond solar evaporation [ref]	0.0147
$CaCO_3$	**Lab-scale:** 1. Seed crystallization [44] **Full-scale:** 1. Chemical precipitation using lime dosing [ref]	0.124
MgO, $MgCO_3$, $Mg(OH)_2$, $MgSO_4$	**Lab-scale:** 1. Chemical precipitation [38] 2. MD/MDC 3. ED followed by NaOH precipitation 4. EFC **Pilot-scale:** 1. Adsorption by carboxylate functionalized ion-exchange resin followed by desorption using Na_2CO_3 and $NaHCO_3$ **Full-scale:** 1. Chemical precipitation using lime/dolomite	0.435 (price as MgO basis)
NaCl	**Lab-scale:** 1. MD/MDC 2. ED followed by evaporation – crystallization 3. EFC **Full-scale:** 1. Solar evaporation ponds	0.049
K_2O, KCl	**Lab-scale:** 1. Selective ion-exchange using zeolite 2.EFC	0.593 (price as K_2O basis)
Na_2SO_4	**Lab-scale:** 1. EFC **Full-scale:** 1. SAL-PROC	0.0157
Li	**Lab-scale:** 1. Adsorption using MnO_2 Pilot-scale: 1. Adsorption on granulated Mn oxide	4.54
Rb	**Lab-scale**: 1. Batch and column adsorption on Cstreat	84,611
Cs	**Lab-scale:** 1. Batch and column adsorption on Cstreat **Full-scale:** 1. Liquid-liquid extraction by Calixarenes	76,400
Sr	**Lab-scale:** 1. Adsorption on zeolite, alginate microsphere and titanate nanotubes	31.5
U	**Lab-scale:** 1. Adsorption on electrospun nanofibers, polyacrylonitrile/polygorksite, and porous polymer all functionalized with amidoxime group	77 – 120#

* market price data from United States Geological Survey (USGS) [83]; # U market price data from International Atomic Energy Agency (IAEA)/The Nuclear Energy Agency (NEA) report 2018 [85]

Electrodialysis is another membrane-based separation scheme where ions migrate across semipermeable membranes (anion-selective for anions to pass through and cation-selective for cations to pass through) by applying an external electric field. The anion and cation permeable membranes are oriented facing each other, and a unidirectional electric field perpendicular to the membrane surface is applied between them. The membranes can be arranged in a series fashion with alternating anion and cation-selective membranes, and the assembly is placed between an anode and a cathode to prepare an ED cell. The migration of anions and cations across membranes leads to concentration and depletion of ions in the respective cells, ultimately leading to desalting and salt concentration. The development of selective monovalent cationic and anionic membrane has been instrumental in the application of ED as a scheme for RO brine treatment, since it separates monovalent ions like Na^+ and Cl^- from divalent ions like Ca^{2+}, Mg^{2+}, CO_3^{2-}, and SO_4^{2-}, and thus makes crystallization of salts like NaCl possible [16, 33]. ED increases the salt concentration from 0.2 – 2% to 12 – 20% with an energy consumption of 1 – 7 kWh/m^3 as compared to 25 kWh/m^3 required for thermal evaporation [10].

Most of the reported studies in ED have been carried out for NaCl crystallization. ED has also been used as a pretreatment step to remove Ca^{2+} and SO_4^{2-} ions in order to prevent precipitation of gypsum during evaporative cooling [74]. Apart from NaCl, ED can also be utilized to yield products like Mg $(OH)_2$ and Br_2 gas, since, after NaCl recovery, bromide-rich bittern can be treated with Cl_2 gas to oxidize Br^- to Br_2 gas. The remaining solution after NaCl removal is concentrated by 5 times for Mg^{2+} ions, which can be precipitated as $Mg(OH)_2$ [16]. Fouling caused by organic and colloidal components must be taken into account before designing an ED process. Ca must be removed by pretreating with Na_2CO_3, since carbonate and sulfate salts of Ca cause a significant scaling issue. ED is in a developing stage and synthesis of ion-specific membranes can make it as a viable tool for targeted resource recovery, e.g., Li-ion specific membrane was developed by Hoshino, which arrests the remaining ions but passes only Li ions [75]. Further development in this direction can make ED an attractive method for resource recovery from RO concentrate.

3.2.5. Other Miscellaneous Methods

Apart from the methods discussed above, there are many studies that have been reported on RO concentrate treatment with the aim of resource recovery. These methods have shown promise and can be explored in the future for salts recovery from brine reject. Abdul-Wahab and Al-Weshahi (2009) [30] using electrolysis demonstrated the feasibility of recovering sodium hypochlorite from brine reject. Sodium hypochlorite is anti-foulant and a disinfectant, therefore, recovering it can cut down the disinfection cost and can potentially bring down the cost of produced water. Mohite & Khopkar (1985) [76] achieved more than 80% recovery of Rb using a liquid-liquid extraction method where BAMBP [4-tertiary-butyl-2-(a-methyl benzyl) phenol] or 4-sec-butyl-2-(a-methyl benzyl) phenol was used as an extractant. Though the study recovered Rb, the reagent was not specific for Rb. The other alkali-metal ions including of Cs were co-extracted. In another experiment reported in the study, a Dowex 50W-X8 cation exchange resin (H^+ - form) column was utilized, where the metal ions were eluted in a sequential manner using different eluents. Rb was eluted using 2M hydrochloric acid and subsequently extracted by BAMBP. Sequential selective liquid-liquid extraction has also been applied to recover Cs (using Calixarenes), In, Ga (both together using three organic acids), and Rb (using titanium silicates based cation exchange resins) from the reject brine at a RO desalination plant in Skhira, Tunisia apart from other recovery techniques such as evaporative crystallization, adsorption, etc. [77]. Another method of crystallization is eutectic freeze point crystallization (EFC) which involves achieving eutectic temperature to separate the aqueous solution into pure water and solid salts. It is because the heat of fusion of water is significantly lower than the heat of evaporation of water, roughly six-time lower, therefore it is far easier to freeze the water as ice than to evaporate it as steam and separate pure water and solid salts. Therefore, it is concluded in a study which compared salt production and energy cost for EFC and evaporative crystallization that EFC process yields 80 – 85% energy savings than evaporative crystallization [78]. However, the capital cost for a 100 m^3/day plant came out to be as high as $ 1.62 million in 2009, mainly due to the complex mechanical equipment needed for separation and product washing [79]. Randall et al. (2011) applied EFC to sequentially recover calcium sulfate, sodium sulfate and potable water from industrial brine with a 97% reduction in waste disposal [80]. The key challenge with EFC is the separation of individual salts from the solid phase. Some biological methods have also been demonstrated for treating/utilizing reject brine to develop algal pond and wetlands. The development of wetlands can provide ecological habitat to many plants and organisms and can be useful in agriculture, and industrial development in areas prone to water scarcity and drought. RO concentrate can be used to cultivate many plants, animals, and micro-organisms. Most studies have focused on cultivating Artemia (brine shrimp) and microalgae *Dunaliella Salina* [81, 33]. Artemia is a crustacean which thrives in high salinity water (70 – 340 g/L) which can be used as a food source for larvae and some species of fish and shellfish [81]. *Dunaliella Salina* is a Chlorophyta which can store carotenoids within its tissue (up to 14% of its dry weight) and has high amounts of glycerol within its biomass. This alga can be harvested for useful food and drugs application. Studies have shown promising results with this algae, and further investigation is needed to fully understand the carotenoid production within its cells. Moreover, algal strains having higher glycerol content can also be explored for biodiesel production [82]. Other algae that have been studied to reduce TDS from desalination brines include *Scendesmus* algae, and *Chlorella* [33]. These studies have shown high TDS removal from 13% on day 1 to 63% on day 6 by treating the RO concentrate with TDS ranging from 40 g/L to 80 g/L [83]. Some grass-like *Paspalum* can also be grown and irrigated with reject brine [33]. However, more research is needed in this direction before it can be fully implemented.

In Table 2, key results, advantages, and limitations from the application of different methodologies have been summarized. Table 3 shows the minerals which have been successfully recovered, the scale at which they were recovered and their market price.

Conclusion

RO desalination is considered crucial for increasing potable water availability. The resulting RO concentrate discharged during RO desalination is considered harmful if released into the environment. Therefore, several studies have been reported in recent years that discuss the feasibility of extracting valuable resources and minerals and methods of treating RO discharge brines. The key methods include chemical precipitation, crystallization, evaporation, membrane-based methods such as membrane distillation crystallization and electrodialysis, adsorption and ion-exchange processes. Chemical precipitation, crystallization, and membrane-based methods are useful in the case where recovery is sought for high ionic strength minerals/salts, whereas adsorption/ion-exchange can be useful for trace minerals recovery. To achieve a holistic recovery model, process integration is necessary. The biological methods utilizing algae can be a useful alternative as it fixes the minerals in the tissue of algae which can be recovered in other usable forms such as biodiesel or food. Moreover, it leads to the generation of wetlands which can provide water to nearby areas for agriculture and other relevant needs. Depending on the environmental, social and economic constraints, the above methods can be judiciously selected. However, a key question is of integrating any resource recovery technology within an active facility, which would then require overall plant retrofitting and process modifications. Presently, most RO desalination units discharge RO concentrates in nearby sea/ocean if they are operating near shores, due to cost considerations. This is surely cheaper but may prove to be unsustainable in the long run, given many studies point in the direction of environmental harms and the possibility of resource recovery and improving water recovery overall.

References

[1] Guppy, Lisa; Kelsey, Anderson. *Water Crisis Report*. United Nations University Institute for Water, Environment and Health, Hamilton, Canada, 2017. https://inweh.unu.edu/wp-content/uploads/2017/11/Global-Water-Crisis-The-Facts.pdf [Accessed on 04/08/2019].

[2] World Health Organization, and UNICEF. "*Progress on drinking water, sanitation and hygiene: 2017 update and SDG baselines.*," (2017).

[3] Yang, Hong; Peter, Reichert; Karim, C. Abbaspour; Alexander, JB Zehnder. "*A water resources threshold and its implications for food security.*," (2003), 3048-3054.

[4] Water, UN. "*The United Nations World Water Development Report 2017: wastewater the untapped resource.*," (2017).

[5] WERF. Nutrient recovery state of the knowledge. Water environment research foundation, USA, 2010. Available at: www.werf.org/c/2011Challenges/Nutrient_Recovery.aspx.
[Accessed on 04/08/2019].

[6] Elimelech, Menachem; William, A. Phillip. "The future of seawater desalination: energy, technology, and the environment." *Science*, 333, no. 6043, (2011), 712-717.

[7] Roberts, David A; Emma, L. Johnston; Nathan, A. Knott. "Impacts of desalination plant discharges on the marine environment: A critical review of published studies." *Water research*, 44, no. 18, (2010), 5117-5128.

[8] Greenlee, Lauren, F; Desmond, F. Lawler; Benny, D. Freeman; Benoit, Marrot; Philippe, Moulin. "Reverse osmosis desalination: water sources, technology, and today's challenges." *Water research*, 43, no. 9, (2009), 2317-2348.

[9] Miller, James E. "Review of water resources and desalination technologies." *Sandia national labs unlimited release report SAND-2003-0800*, (2003).

[10] Pérez-González, A; Urtiaga, AM; Ibáñez, R; Ortiz, I. "State of the art and review on the treatment technologies of water reverse osmosis concentrates." *Water research*, 46, no. 2, (2012), 267-283.

[11] Meneses, Montse; Jorgelina, C. Pasqualino; Raquel, Céspedes-Sánchez; Francesc, Castells. "Alternatives for reducing the environmental impact of the main residue from a desalination plant." *Journal of Industrial Ecology*, *14*, no. 3, (2010), 512-527.

[12] Voutchkov, Nikolay. "Overview of seawater concentrate disposal alternatives." *Desalination*, *273*, no. 1, (2011), 205-219.

[13] Mohamed, AMO; Maraqa, M; Al Handhaly, J. "Impact of land disposal of reject brine from desalination plants on soil and groundwater." *Desalination*, 182, no. 1-3, (2005), 411-433.

[14] Subramani, Arun; Joseph, G. Jacangelo. "Treatment technologies for reverse osmosis concentrate volume minimization: A review." *Separation and Purification Technology*, 122, (2014), 472-489.

[15] Shahmansouri, Arash; Joon, Min; Liyan, Jin; Christopher, Bellona. "Feasibility of extracting valuable minerals from desalination concentrate: a comprehensive literature review." *Journal of Cleaner Production*, 100, (2015), 4-16.

[16] Loganathan, Paripurnanda; Gayathri, Naidu, Saravanamuthu, Vigneswaran. "Mining valuable minerals from seawater: a critical review." *Environmental Science: Water Research & Technology, 3*, no. 1, (2017), 37-53.

[17] Bird, R. Byron; Warren, E. Stewart; Edwin, N. Lightfoot. *Transport phenomena*. John Wiley & Sons, 2007.

[18] Lonsdale, HK; Ulrich, Merten; Riley, RL. "Transport properties of cellulose acetate osmotic membranes." *Journal of Applied Polymer Science*, 9, no. 4, (1965), 1341-1362.

[19] Paul, Donald R. "Reformulation of the solution-diffusion theory of reverse osmosis." *Journal of Membrane Science*, 241, no. 2, (2004), 371-386.

[20] Lewis, Gilbert Newton. "The osmotic pressure of concentrated solutions, and the laws of the perfect solution." *Journal of the American Chemical Society*, 30, no. 5, (1908), 668-683.

[21] Green, Don W; Robert, H. Perry. *Perry's Chemical Engineers' Handbook/edición Don W. Green y Robert H. Perry*. No. C 660.28 P47 2008. 1997.

[22] Sanza, Miguel Angel; Véronique, Bonnélyea; Gerardo, Cremerb. "Fujairah reverse osmosis plant: 2 years of operation." *Desalination*, 203, no. 1-3, (2007), 91-99.

[23] Tang, Chuyang Y; Zhe, Yang; Hao, Guo; Jason, J. Wen; Long, D. Nghiem; Emile, Cornelissen. "*Potable water reuse through advanced membrane technology.*," (2018), 10215-10223.

[24] Lau, WJ; Ismail, AF; Misdan, N; Kassim, MA. "A recent progress in thin film composite membrane: a review." *Desalination*, 287, (2012), 190-199.

[25] Lau, WJ; Stephen, Gray; Matsuura, T; Emadzadeh, D; Paul Chen, J; Ismail, AF. "A review on polyamide thin film nanocomposite (TFN) membranes: history, applications, challenges and approaches." *Water research*, 80, (2015), 306-324.

[26] Werber, Jay R; Chinedum, O. Osuji; Menachem, Elimelech. "Materials for next-generation desalination and water purification membranes." *Nature Reviews Materials*, 1, no. 5, (2016), 16018.

[27] Choi, Yong-Jun; June-Seok, Choi; Hyun-Je, Oh; Sangho, Lee; Dae, Ryook Yang; Joon, Ha Kim. "Toward a combined system of forward osmosis and reverse osmosis for seawater desalination." *Desalination*, 247, no. 1-3, (2009), 239-246.

[28] Curcio, Efrem; Enrico, Drioli. "Membrane distillation and related operations—a review." *Separation and Purification Reviews*, 34, no. 1, (2005), 35-86.

[29] Abdulsalam, Alrowaished; Azni, Idris; Thamer, A. Mohamed, and Amimul Ahsan. "An integrated technique using solar and evaporation ponds for effective brine disposal management." *International Journal of Sustainable Energy*, 36, no. 9, (2017), 914-925.

[30] Abdul-Wahab, Sabah A; Mohammed, A. Al-Weshahi. "Brine management: substituting chlorine with on-site produced sodium hypochlorite for environmentally improved desalination processes." *Water resources management*, 23, no. 12, (2009), 2437-2454.

[31] Pal, Sangita; Prabhakar, S; Thalor, KL; Tewari, PK. "Strategy of deriving' wealth from waste' from concentrated brine of desalination plant." *International Journal of Nuclear Desalination*, 4, no. 2, (2010), 189-197.

[32] Sorour, Mohamed H; Heba, A. Hani; Hayam, F. Shaalan; Ghada, A. Al-Bazedi. "Schemes for salt recovery from seawater and RO brines using chemical precipitation." *Desalination and Water Treatment*, 55, no. 9, (2015), 2398-2407.

[33] Giwa, A; Dufour, V; Al Marzooqi, F; Al Kaabi, M; Hasan, SW. "Brine management methods: Recent innovations and current status." *Desalination*, 407, (2017), 1-23.

[34] Rahardianto, Anditya; Brian, C. McCool; Yoram, Cohen. "Accelerated desupersaturation of reverse osmosis concentrate by chemically-enhanced seeded precipitation." *Desalination*, 264, no. 3, (2010), 256-267.

[35] Hofs, B; et al. "*Increased sustainability of softening by producing pure calcite pellets for reuse.*," (2015).

[36] Kim, Myung-man; James, Au; Anditya, Rahardianto; Julius, Glater; Yoram, Cohen; Fredrick, W. Gerringer; Christopher, J. Gabelich. "Impact of conventional water treatment coagulants on mineral scaling in RO desalting of brackish water." *Industrial & Engineering Chemistry Research*, 48, no. 6, (2009), 3126-3135.

[37] Shih, Wen-Yi; Junbo, Gao; Anditya, Rahardianto; Julius, Glater; Yoram, Cohen; Christopher, J. Gabelich. "Ranking of antiscalant performance for gypsum scale suppression in the presence of residual aluminum." *Desalination*, 196, no. 1-3, (2006), 280-292.

[38] Xiong, Rihua; Quan, Chen; Jie, Liu; Chang, Wei. "Experimental study on seeded precipitation assisted reverse osmosis for industrial wastewater reuse." *Journal of water process engineering*, 20, (2017), 78-83.

[39] Pervov, AG. "Precipitation of calcium carbonate in reverse osmosis retentate flow by means of seeded techniques—A tool to increase recovery." *Desalination*, 368, (2015), 140-151.

[40] Azadi Aghdam, Mojtaba. "*A Novel Brine Precipitation Process with the Aim of Higher Permeate Water Recovery.*," (2016).

[41] van Houwelingen, Gerard; Rick, Bond; Tom, Seacord; Eric, Fessler. "Experiences with pellet reactor softening as pretreatment for inland desalination in the USA." *Desalination and Water Treatment*, 13, no. 1-3, (2010), 259-266.

[42] Helfferich, Friedrich G. *Ion exchange*. Courier Corporation, 1995.

[43] Kyzas, George; Margaritis, Kostoglou. "Green adsorbents for wastewaters: a critical review." *Materials*, 7, no. 1, (2014), 333-364.

[44] Nishihama, Syouhei; Kenta, Onishi; Kazuharu, Yoshizuka. "Selective recovery process of lithium from seawater using integrated ion exchange methods." *Solvent Extraction and Ion Exchange*, 29, no. 3, (2011), 421-431.

[45] Nakazawa, Naoki; Masao, Tamada; Noriaki, Seko; Kenta, Ooi; Satoshi, Akagawa. "Experimental studies on rare metal collection from seawater." In *Ninth ISOPE Ocean Mining Symposium*. International Society of Offshore and Polar Engineers, 2011.

[46] Chitrakar, Ramesh; Hirofumi, Kanoh; Yoshitaka, Miyai; Kenta, Ooi. "Recovery of lithium from seawater using manganese oxide adsorbent (H1. 6Mn1. 6O4) derived from Li1. 6Mn1. 6O4." *Industrial & Engineering Chemistry Research*, 40, no. 9, (2001), 2054-2058.

[47] Gibert, Oriol; César, Valderrama; Michaela, Peterkóva; José, Luis Cortina. "Evaluation of selective sorbents for the extraction of valuable metal ions (Cs, Rb, Li, U) from reverse osmosis rejected brine." *Solvent Extraction and Ion Exchange*, 28, no. 4, (2010), 543-562.

[48] Carboni, Michaël; Carter, W. Abney; Shubin, Liu; Wenbin, Lin. "Highly porous and stable metal–organic frameworks for uranium extraction." *Chemical Science*, 4, no. 6, (2013), 2396-2402.

[49] Peterskovâ, Michaela; César, Valderrama; Oriol, Gibert; José, Luis Cortina. "Extraction of valuable metal ions (Cs, Rb, Li, U) from reverse osmosis concentrate using selective sorbents." *Desalination*, 286, (2012), 316-323.

[50] Sorour, Mohamed H; Heba, A. Hani; Hayam, F. Shaalan; Marwa, M. El Sayed; Mayyada, MH El-Sayed. "Softening of seawater and desalination brines using grafted polysaccharide hydrogels." *Desalination and Water Treatment*, 55, no. 9, (2015), 2389-2397.

[51] Hong, Hye-Jin; Jungho, Ryu; In-Su, Park; Taegong, Ryu; Kang-Sup, Chung; Byuong-Gyu, Kim. "Investigation of the strontium (Sr (II)) adsorption of an alginate microsphere as a low-cost adsorbent for removal and recovery from seawater." *Journal of environmental management*, 165, (2016), 263-270.

[52] Brus, Jiri; Martina, Urbanova; Jiri, Czernek; Miroslava, Pavelkova; Katerina, Kubova; Jakub, Vyslouzil; Sabina, Abbrent; et al. "Structure and dynamics of alginate gels cross-linked by polyvalent ions probed via solid state NMR spectroscopy." *Biomacromolecules*, 18, no. 8, (2017), 2478-2488.

[53] Hong, Hye-Jin; Jungho, Ryu; In-Su, Park; Taegong, Ryu; Kang-Sup, Chung; Byuong-Gyu, Kim. "Investigation of the strontium (Sr (II)) adsorption of an alginate microsphere as a low-cost adsorbent for removal and recovery from seawater." *Journal of environmental management*, 165, (2016), 263-270.

[54] Ryu, Jungho; Soonhyun, Kim; Hye-Jin, Hong; Jeongsik, Hong; Minsun, Kim; Taegong, Ryu; In-Su, Park; Kang-Sup, Chung; Jum, Suk Jang; Byoung-Gyu, Kim. "Strontium ion (Sr2+) separation from seawater by hydrothermally structured titanate nanotubes: Removal vs. recovery." *Chemical Engineering Journal*, 304, (2016), 503-510.

[55] Weast, Robert C; Melvin, J. Astle; William, H. Beyer. *CRC handbook of chemistry and physics*., Vol. 69. Boca Raton, FL: CRC press, 1988.

[56] Naidu, Gayathri; Tanjina, Nur; Paripurnanda, Loganathan; Jaya. Kandasamy; Saravanamuthu, Vigneswaran. "Selective sorption of rubidium by potassium cobalt hexacyanoferrate." *Separation and Purification Technology*, 163, (2016), 238-246.

[57] Naidu, Gayathri; Paripurnanda, Loganathan; Sanghyung, Jeong; Md Abu, Hasan Johir; Vu, Hien Phuong To; Jaya, Kandasamy; Saravanamuthu, Vigneswaran. "Rubidium extraction using an organic polymer encapsulated potassium copper hexacyanoferrate sorbent." *Chemical Engineering Journal*, 306, (2016), 31-42.

[58] Hajbi, Ferid; Halim, Hammi; Adel, M'nif. "Reuse of RO desalination plant reject brine." *Journal of phase equilibria and diffusion*, 31, no. 4, (2010), 341-347.

[59] Ahmed, Mushtaque; Aro, Arakel; David, Hoey; Muralee, R. Thumarukudy; Mattheus, FA Goosen; Mansour, Al-Haddabi; Abdullah, Al-Belushi. "Feasibility of salt production from inland RO desalination plant reject brine: a case study." *Desalination*, 158, no. 1-3, (2003), 109-117.

[60] Ahmed, Mushtaque; Walid, H. Shayya; David, Hoey; Arun, Mahendran; Richard, Morris; Juma, Al-Handaly. "Use of evaporation ponds for brine disposal in desalination plants." *Desalination*, 130, no. 2, (2000), 155-168.

[61] Abdulsalam, Alrowaished; Azni, Idris; Thamer, A. Mohamed; Amimul, Ahsan. "An integrated technique using solar and evaporation ponds for effective brine disposal management." *International Journal of Sustainable Energy*, 36, no. 9, (2017), 914-925.

[62] Macedonio, F; Katzir, L; Geisma, N; Simone, S; Drioli, E; Gilron, J. "Wind-Aided Intensified eVaporation (WAIV) and Membrane Crystallizer (MCr) integrated brackish water desalination process: Advantages and drawbacks." *Desalination*, 273, no. 1, (2011), 127-135.

[63] Pérez-González, Antía; Ibáñez, R; Gómez, P; Urtiaga, AM; Ortiz, I; Irabien, JA. "Nanofiltration separation of polyvalent and monovalent anions in desalination brines." *Journal of membrane science*, 473, (2015), 16-27.

[64] Urtiaga, AM; Gorri, ED; Ruiz, G; Ortiz, I. "Parallelism and differences of pervaporation and vacuum membrane distillation in the removal of VOCs from aqueous streams." *Separation and Purification Technology*, 22, (2001), 327-337.

[65] Charcosset, Catherine; Roland, Kieffer; Denis, Mangin; François, Puel. "Coupling between membrane processes and crystallization operations." *Industrial & engineering chemistry research*, 49, no. 12, (2010), 5489-5495.

[66] Macedonio, Francesca; Enrico, Drioli. "Hydrophobic membranes for salts recovery from desalination plants." *Desalination and Water Treatment*, 18, no. 1-3, (2010), 224-234.

[67] Curcio, Efrem; Alessandra, Criscuoli; Enrico, Drioli. "Membrane crystallizers." *Industrial & Engineering Chemistry Research*, 40, no. 12, (2001), 2679-2684.

[68] Macedonio, Francesca; Cejna, Anna Quist-Jensen; Al-Harbi, O; Alromaih, H; Al-Jlil, SA; Al Shabouna, F; Enrico, Drioli.

"Thermodynamic modeling of brine and its use in membrane crystallizer." *Desalination*, 323, (2013), 83-92.

[69] Quist-Jensen, Cejna Anna; Francesca, Macedonio; Enrico, Drioli. "Membrane crystallization for salts recovery from brine—an experimental and theoretical analysis." *Desalination and Water Treatment*, 57, no. 16, (2016), 7593-7603.

[70] Drioli, Enrico; Efrem, Curcio; Alessandra, Criscuoli; Gianluca, Di Profio. "Integrated system for recovery of CaCO3, NaCl and MgSO4·7H2O from nanofiltration retentate." *Journal of Membrane Science*, 239, no. 1, (2004), 27-38.

[71] Macedonio, Francesca; Enrico, Drioli. "Hydrophobic membranes for salts recovery from desalination plants." *Desalination and Water Treatment*, 18, no. 1-3, (2010), 224-234.

[72] Drioli, Enrico; Alessandra, Criscuoli; Efrem, Curcio. "Integrated membrane operations for seawater desalination." *Desalination*, 147, no. 1-3, (2002), 77-81.

[73] Al Bazedi, Ghada; Reem, S. Ettouney; Shadia, R. Tewfik; Mohamed, H. Sorour; Mahmoud, A. El-Rifai. "Salt recovery from brine generated by large-scale seawater desalination plants." *Desalination and Water Treatment*, 52, no. 25-27, (2014), 4689-4697.

[74] Tanaka, Yoshinobu; Reo, Ehara; Sigeru, Itoi; Totaro, Goto. "Ion-exchange membrane electrodialytic salt production using brine discharged from a reverse osmosis seawater desalination plant." *Journal of Membrane Science*, 222, no. 1-2, (2003), 71-86.

[75] Hoshino, Tsuyoshi. "Innovative lithium recovery technique from seawater by using world-first dialysis with a lithium ionic superconductor." *Desalination*, 359, (2015), 59-63.

[76] Mohite, BS; Khopkar, SM. "Solvent extraction separation of rubidium with dicyclohexano-18-crown-6." *Talanta*, 32.7, (1985), 565-567.

[77] Le Dirach, J; Nisan, S; Poletiko, C. "Extraction of strategic materials from the concentrated brine rejected by integrated nuclear desalination systems." *Desalination*, 182, no. 1-3, (2005), 449-460.

[78] Nathoo, J; Jivanji, R; Lewis, AE. "Freezing your brines off: eutectic freeze crystallization for brine treatment." In *International mine water conference*, pp. 431-437, 2009.

[79] Randall, DG; Nathoo, J. "A succinct review of the treatment of Reverse Osmosis brines using Freeze Crystallization." *Journal of water process engineering*, 8, (2015), 186-194.

[80] Randall, DG; Nathoo, J; Lewis, AE. "A case study for treating a reverse osmosis brine using Eutectic Freeze Crystallization—Approaching a zero waste process." *Desalination*, 266, 1-3, (2011), 256-262.

[81] Zhuang, Xuliang; Zhen, Han; Zhihui, Bai; Guoqiang, Zhuang; Hojae, Shim. "Progress in decontamination by halophilic microorganisms in saline wastewater and soil." *Environmental Pollution*, 158, no. 5, (2010), 1119-1126.

[82] Raja, R; Hemaiswarya, S; Rengasamy, R. "Exploitation of Dunaliella for β-carotene production." *Applied Microbiology and Biotechnology*, 74, no. 3, (2007), 517-523.

[83] El Sergany, R; El Fadly, M; El Nadi, A. "Brine desalination by using algae ponds under nature conditions." *Am. J. Environ. Eng*, 4, (2014), 75-79.

[84] National Minerals Information Center. Historical statistics for mineral and material commodities in the United States. 2014 version, United States Geological Survey. Available at:
[Accessed on 09/17/2019].

[85] Uranium 2018: Resources, Production and Demand. A joint report by the Nuclear Energy Agency and the International Atomic Energy Agency. Copyright OECD 2018. Available at: https://www.oecd-nea.org/ndd/pubs/2018/7413-uranium-2018.pdf
[Accessed on: 09/17/2019].

EDITORS' CONTACT INFORMATION

Veera Gnaneswar Gude, PhD

Civil & Environmental Engineering Department
Mississippi State, MS, USA
Email Address: Gude@cee.msstate.edu

Venkataramana Gadhamshetty, PhD

Department of Environmental and Civil Engineering
South Dakota School of Mines and Technology,
Rapid City, South Dakota, US
Email Address: venkata.gadhamshetty@sdsmt.edu

Ramanitharan Kandiah, PhD

Professor, Water Resources Management & Environmental Engineering
International Center for Water Resources Management
Central State University, Wilberforce, OH, USA
Email Address: rkandiah@centralstate.edu

INDEX

C

D

E

F

G

H

I

J

K

L

M

N

O

P

Q

R

T

U

V

W

Y

Z

WATER RESOURCE PLANNING, DEVELOPMENT AND MANAGEMENT

SUSTAINABLE WATER

RESOURCES, MANAGEMENT AND CHALLENGES

Water Resource Planning, Development and Management

Additional books and e-books in this series can be found on Nova's website under the Series tab.

WATER RESOURCE PLANNING, DEVELOPMENT AND MANAGEMENT

SUSTAINABLE WATER

RESOURCES, MANAGEMENT AND CHALLENGES

VEERA GNANESWAR GUDE, PHD
VENKATARAMANA GADHAMSHETTY, PHD
AND
RAMANITHARAN KANDIAH, PHD
EDITORS

nova
science publishers
New York

Library of Congress Cataloging-in-Publication Data

ISBN: 978-1-53617-335-2

Published by Nova Science Publishers, Inc. † New York